Multi-field Modeling of Soft Active Materials

Properties and Design

Rui Xiao
China

WILEY-VCH

Author

Rui Xiao
Zhejiang University
No. 38, Zheda Road
Xihu District
Hangzhou
China

Cover Design: Wiley

Cover Image: Courtesy of Rui Xiao

Typesetting: Lumina Datamatics

Printing and Binding: CPI Group (UK) Ltd, Croydon, CR0 4YY

Library of Congress Card No.: applied for

British Library Cataloguing-in-Publication Data
A catalogue record for this book is available from the British Library.

Bibliographic information published by the Deutsche Nationalbibliothek
The Deutsche Nationalbibliothek lists this publication in the Deutsche Nationalbibliografie; detailed bibliographic data are available on the Internet at http://dnb.d-nb.de.

Print ISBN: 9783527354689
ePDF ISBN: 9783527850396
ePub ISBN: 9783527850389
oBook ISBN: 9783527850402

C9783527354689_240726

Contents

Foreword *vii*
Preface *ix*
Acknowledgments *xi*

1 Basics of Continuum Mechanics *1*
1.1 Vectors and Tensors *1*
1.1.1 Vector *1*
1.1.2 Index Notation *2*
1.1.3 Tensor *3*
1.1.4 Gradient, Divergence, and Curl *4*
1.2 Kinematics *5*
1.2.1 Deformation Gradient *5*
1.2.2 Strain Tensor *7*
1.3 Stress *8*
1.4 Balance Principles *9*
1.4.1 Material Derivative and Spatial Derivative *10*
1.4.2 Reynolds Transport Theorem *10*
1.4.3 Conservation of Mass *11*
1.4.4 Balance of Momentum *11*
1.4.5 Balance of Angular Momentum *12*
1.4.6 Balance of Mechanical Energy *13*
1.4.7 Balance of Energy *14*
1.4.8 Entropy Inequality *15*
References *17*

2 Hyperelastic, Viscoelastic, and Damage Models *19*
2.1 Introduction *19*
2.2 Hyperelastic Models *21*
2.2.1 Constitutive Equations of Hyperelastic Materials *21*
2.2.2 Hyperelastic Models *24*

2.2.3 Results *29*
2.3 Viscoelastic Models *31*
2.3.1 One-dimensional Small Strain Viscoelastic Models *31*
2.3.2 Finite Deformation Viscoelastic Models *33*
2.3.3 Results *38*
2.3.4 Finite Element Simulation *39*
2.4 Damage Models *41*
2.4.1 Continuum Damage Model *41*
2.4.2 Network Alteration Theory *42*
2.4.3 Progressively Damage Model *42*
2.4.4 Extended Network Alteration Theory *43*
2.4.5 Extended Progressively Damage Model *45*
2.4.6 Results *47*
2.5 Conclusion *51*
References *53*

3 A Thermomechanical Coupled Model for Amorphous Shape-memory Polymers *57*
3.1 Introduction *57*
3.2 Thermodynamics *59*
3.2.1 Kinematics *59*
3.2.2 Thermodynamic Framework *60*
3.2.3 Constitutive Relationships *64*
3.2.4 A Reduced Version of the Constitutive Model *67*
3.3 Parameter Determination *68*
3.4 Results *68*
3.4.1 Performance of the Viscoelastic Model *68*
3.4.2 Performance of the Effective Temperature Model *71*
3.5 Discussion *73*
3.6 Conclusion *74*
References *75*

4 An Electromechanical Coupled Model for Dielectric Elastomers *79*
4.1 Introduction *79*
4.2 Theory *80*
4.2.1 Thermodynamic Framework *81*
4.2.2 An Electro-hyperelastic Model *84*
4.2.3 An Electro-hyper-viscoelastic Model *86*
4.2.4 Electromechanical Instability of Dielectric Elastomers *86*
4.3 Finite Element Implementation *90*
4.3.1 Constitutive Relation *90*
4.3.2 Weak Forms *91*
4.4 Conclusion *94*
References *95*

5 A Magnetomechanical Coupled Model for Magnetoactive Soft Materials *99*
5.1 Introduction *99*
5.2 A Magnetomechanical Coupled Model for h-MREs *101*
5.2.1 Kinematics *101*
5.2.2 F-based Model *102*
5.2.3 R-based Model *103*
5.2.4 Comparing R-based and F-based Models *104*
5.3 Magnetic Activated Shape-memory Polymers *105*
5.3.1 Constitutive Theory *106*
5.3.2 Parameter Determination *109*
5.3.3 Simulation Results *111*
5.4 Conclusion *112*
Appendix *113*
References *114*

6 Multi-field Modeling of Liquid Crystal Elastomers *117*
6.1 Introduction *117*
6.2 General Theory for Liquid Crystal Elastomers *119*
6.2.1 Liquid Crystal *119*
6.2.2 Nematic Polymers *121*
6.2.3 Neoclassical Model *122*
6.3 Thermomechanical Coupled Model for Monodomain Liquid Crystal Elastomers *123*
6.4 A Viscoelastic Micropolar Theory for Monodomain Liquid Crystal Elastomers *125*
6.4.1 Kinematics *126*
6.4.2 Balance of Linear and Angular Momentum *126*
6.4.3 Balance of Energy and the Second Law of Thermodynamics *127*
6.4.4 Constitutive Equations *128*
6.4.5 Simplification for the Non-gradient Case *130*
6.4.6 Free Energy Density *132*
6.5 Theory for Polydomain Liquid Crystal Elastomers *134*
6.6 Conclusion *139*
References *140*

7 A Chemomechanical Model for Hydrogels *143*
7.1 Introduction *143*
7.2 Thermodynamics *145*
7.3 Constitutive Relations *148*
7.3.1 Neutral Gels *148*
7.3.2 Fiber-reinforced Gels *149*
7.3.3 Diffusion in Glassy Elastomers *150*
7.3.4 Numerical Method *152*

7.4 Results *153*
7.4.1 Neutral Gels and Fiber-reinforced Gels *153*
7.4.2 Diffusion in Glassy Polymers *155*
7.5 Conclusion *157*
References *157*

Index *161*

Foreword

One of the major advances in the field of mechanics comes from the emergence of *Mechanics of Soft Materials*. Its fast progress is facilitated by four factors: the finite deformation formulation becomes a necessity at its kinematic extreme; comparable effects are encountered due to mechanical, electrical, and magneto fields to leverage multi-physics coupling; the crossing between solid and fluid media makes its constitutive description dangling from non-Newtonian fluids to weakly shear-resistant solids; and the affinity to biological media leads to the interfacial matching between soft materials and human bodies and the intervention of biochemical processes. It is exactly built on this context; the book that addresses this need, *Multi-field Modeling of Soft Active Materials: Properties and Design,* is brought to the science and education forefront.

Prof. Rui XIAO is an energetic scholar with a destiny to engage in the mission of elucidating soft active materials. Rui got his PhD 10 years ago at Johns Hopkins University, and has been my young colleague at the Department of Engineering Mechanics, Zhejiang University, since 2019. He has been pursuing an academic career with a focus on the constitutive behavior of soft materials, encompassing shape-memory polymers, dielectric elastomers, magnetic-responsive elastomers, liquid crystal elastomers, and hydrogels. Though these contents are all hot topics in research, the current literature is mostly composed of the technical papers, seldom with heavyweight monographs to address comprehensively the mechanical essence of the subject. Rui's monograph has an ambitious goal: to build a knowledge framework on the mechanics characterization of soft active materials, from properties to designs, from mechanics foundation to multi-physics coupling, from shape memory to chemo-mechanical formulation. The methodology runs through continuum mechanics (Chapter 1), hyper-elastic and visco-elastic modeling (Chapter 2), and a spectrum of cases to exemplify multi-field modeling (Chapters 3–7). All formulations start from the thermodynamics framework, then derive constitutive relations, and finally to the finite element implementation. The content covers all major categories of soft materials: amorphous shape-memory polymers (Chapter 3), dielectric elastomers (Chapter 4), magnetoactive soft materials (Chapter 5), liquid crystal elastomers (Chapter 6), and hydrogels (Chapter 7).

With this book as the foundation, the readers may find a common approach to model soft active materials under multi-field situations. And even go beyond, to explore a pathway to construct and to model "intelligent soft materials" with self-healing, self-configuration, self-programmable, and self-growth capabilities.

Wei Yang
Professor of Zhejiang University
Member of Chinese Academy of Science

Preface

Soft active materials, characterized by their capacity for large deformations in response to various fields, are pivotal in the design of smart structures for applications such as soft robotics and biomedical devices. This book presents a comprehensive exploration of the multi-field modeling of soft active materials, encompassing fundamental principles and recent advancements.

My engagement with this field commenced in 2009 during my doctoral studies, where my research centered on shape-memory polymers. At that time, the mechanics of soft matter was a nascent area, with investigations primarily focused on systems like hydrogels and dielectric elastomers. Concurrent with my PhD studies, I also conducted preliminary explorations in these domains. Subsequently, as I transitioned to independent research, the development of responsive soft materials, including magnetoactive elastomers, liquid crystal elastomers, and fiber-reinforced hydrogels, progressed rapidly. While theoretical models were developed to describe individual material systems, a consolidated resource detailing the commonalities in their multi-field coupling theories was absent. This gap motivated me to write this book.

This book addresses this need by providing a coherent theoretical framework applicable to these diverse responsive materials. It is my aspiration that this work will offer valuable insights to researchers and serve as a foundation for the development of models for increasingly complex systems in the future.

The book begins with the basis of continuum mechanics in Chapter 1, introducing vectors, tensors, kinematics, stress, and balance principles. Chapter 2 details constitutive models for hyperelasticity viscoelasticity, and damage. While these topics are typically introduced in a standard *Continuum Mechanics* course, they serve as the foundation of multi-field models developed in the subsequent chapters.

Thermomechanical coupling in amorphous shape-memory polymers (SMPs) is discussed in Chapter 3, featuring the effective temperature model and its application to solvent-driven shape-memory behaviors. Chapter 4 explores electromechanical coupling in dielectric elastomers, including a thermodynamic framework and analysis of electromechanical instability. Chapter 5 presents magnetic-mechanical coupling in magnetoactive soft materials, comparing models for magnetorheological elastomers and incorporating glass transition mechanisms. Multi-field modeling of liquid crystal elastomers (LCEs) is covered in Chapter 6, including theories for monodomain and polydomain LCEs. Finally, Chapter 7 discusses chemo-mechanical modeling of kinetic diffusion in hydrogels, addressing the coupling between solvent diffusion and large deformation.

Overall, this book aims to provide researchers, engineers, and graduate students in mechanical engineering, materials science, and biomedical engineering a robust understanding of soft active materials and their multi-field modeling.

Rui Xiao
Zhejiang University
Hangzhou, China

Acknowledgments

I owe my deepest gratitude to Professor Thao D. Nguyen, with whom I had the privilege of working during my PhD studies at Johns Hopkins University. Her guidance, unwavering support, and exacting standards for research have profoundly influenced not only my doctoral journey but also my subsequent career as an independent researcher.

I was honored to study and collaborate with Professor Paul Steinmann at Friedrich-Alexander University of Erlangen-Nürnberg as a Humboldt Research Fellow. His exceptional expertise and insightful perspectives have inspired me in countless ways.

I am deeply indebted to my esteemed colleagues, Professors Wei Yang, Weiqiu Chen, Shaoxing Qu, Jin Qian, Jizhou Song, Zhe Chen, and Guoyong Mao, for their invaluable collaborations, which have been instrumental in shaping this book. Their contributions were essential to the realization of this work.

Special thanks are due to my dedicated students, especially Liqian Wang, Junwei Xu, Peng Sun, Zengting Xu, Lingrui Zhu, Kunquan Wang, and Le Du, for their meticulous work in preparing figures, reviewing the manuscript, and, most importantly, for the foundational research that directly informs a large portion of this book.

1

Basics of Continuum Mechanics

1.1 Vectors and Tensors

Continuum mechanics relies on vectors and tensors. We first introduce vector/tensor operations. In this book, scalars are denoted by italic lowercase letters, while vectors and tensors are represented by upright boldface letters.

1.1.1 Vector

If a physical quantity contains only magnitude information, such as time, mass, temperature, density, and length, it is referred to as a scalar. A vector represents a directed line segment in space. The length of the line segment indicates the magnitude of the vector, while the direction of the line segment represents the direction of the vector. It is used to describe physical quantities that have both direction and magnitude, such as displacement, velocity, acceleration, and force.

The addition of vectors follows the same rules as scalar addition, satisfying the commutative law and the associative law of addition:

$$\mathbf{u}+\mathbf{v}=\mathbf{v}+\mathbf{u}, \tag{1.1a}$$

$$(\mathbf{u}+\mathbf{v})+\mathbf{w}=\mathbf{u}+(\mathbf{v}+\mathbf{w}). \tag{1.1b}$$

The sum of two vectors is still a vector, and the magnitude and direction of the resultant vector are determined according to the parallelogram law.

The product of a vector $\mathbf{u}$ and a scalar α is a new vector that has the same direction as $\mathbf{u}$ if $\alpha > 0$, or the opposite direction if $\alpha < 0$. The multiplication of a scalar and a vector satisfies the commutative law, the associative law, and the distributive law as follows:

$$\alpha\mathbf{u}=\mathbf{u}\alpha, \tag{1.2a}$$

$$(\alpha\beta)\mathbf{u}=\alpha(\beta\mathbf{u}), \tag{1.2b}$$

$$(\alpha+\beta)\mathbf{u}=\alpha\mathbf{u}+\beta\mathbf{u}, \tag{1.2c}$$

$$\alpha(\mathbf{u}+\mathbf{v})=\alpha\mathbf{u}+\alpha\mathbf{v}. \tag{1.2d}$$

Multi-field Modeling of Soft Active Materials: Properties and Design, First Edition. Rui Xiao.

The dot product of vectors $\mathbf{u}$ and $\mathbf{v}$, denoted by $\mathbf{u}\cdot\mathbf{v}$, results in a scalar:

$$\mathbf{u}\cdot\mathbf{v} = |\mathbf{u}||\mathbf{v}|\cos\theta, \tag{1.3}$$

where θ is the angle between two vectors $\mathbf{u}$ and $\mathbf{v}$. The dot product of vectors has the following properties:

- $\mathbf{u}\cdot\mathbf{v} = \mathbf{v}\cdot\mathbf{u}$,
- $\mathbf{u}\cdot(\mathbf{v}+\mathbf{w}) = \mathbf{u}\cdot\mathbf{v}+\mathbf{u}\cdot\mathbf{w}$.

The cross product of vectors $\mathbf{u}$ and $\mathbf{v}$, denoted by $\mathbf{u}\times\mathbf{v}$, results in a vector. Its magnitude is given by:

$$|\mathbf{u}\times\mathbf{v}| = |\mathbf{u}||\mathbf{v}|\sin\theta. \tag{1.4}$$

The direction of $\mathbf{u}\times\mathbf{v}$ is perpendicular to the plane formed by $\mathbf{u}$ and $\mathbf{v}$, which follows the right-hand rule. The cross product of vectors has the following properties:

- $\mathbf{u}\times\mathbf{v} = -\mathbf{v}\times\mathbf{u}$,
- $\mathbf{u}\times(\mathbf{v}+\mathbf{w}) = \mathbf{u}\times\mathbf{v}+\mathbf{u}\times\mathbf{w}$.

1.1.2 Index Notation

To introduce the coordinate (or component) expression of a vector in a right-handed orthogonal system, we first introduce a set of fixed basis vectors $\mathbf{e}_1, \mathbf{e}_2, \mathbf{e}_3$, called a Cartesian basis, with the following properties:

$$\mathbf{e}_1\cdot\mathbf{e}_2 = \mathbf{e}_1\cdot\mathbf{e}_3 = \mathbf{e}_2\cdot\mathbf{e}_3 = 0, \quad \mathbf{e}_1\cdot\mathbf{e}_1 = \mathbf{e}_2\cdot\mathbf{e}_2 = \mathbf{e}_3\cdot\mathbf{e}_3 = 1. \tag{1.5}$$

These three vectors are of unit length and mutually orthogonal. A vector in three-dimensional space can be expressed as

$$\mathbf{u} = u_1\mathbf{e}_1 + u_2\mathbf{e}_2 + u_3\mathbf{e}_3, \tag{1.6}$$

which can be written in summation notation as

$$\mathbf{u} = \sum_{i=1}^{3} u_i\mathbf{e}_i. \tag{1.7}$$

This expression can be further simplified by using a dummy index to omit the summation symbol, written as

$$\mathbf{u} = u_i\mathbf{e}_i. \tag{1.8}$$

An index that repeats exactly twice in an expression is called a dummy index, which signifies the summation convention. An index that is not summed over in an expression is referred to as a free index. For example:

$$f_i = u_i v_j w_j, \tag{1.9a}$$

$$f_1 = u_1(v_1w_1 + v_2w_2 + v_3w_3), \tag{1.9b}$$

$$f_2 = u_2(v_1w_1 + v_2w_2 + v_3w_3), \tag{1.9c}$$

$$f_3 = u_3(v_1w_1 + v_2w_2 + v_3w_3), \tag{1.9d}$$

where i is the free index and j is the dummy index.

The calculus operations involving scalars and vectors often carry clear physical significance. We begin by introducing two commonly used symbols in continuum mechanics: the Kronecker delta δ_{ij} and the permutation symbol ε_{ijk}. The Kronecker delta δ_{ij} is defined as follows:

$$\mathbf{e}_i \cdot \mathbf{e}_j = \delta_{ij} \equiv \begin{cases} 1, & \text{if} \quad i=j, \\ 0, & \text{if} \quad i \neq j. \end{cases}$$

The following are some of its useful properties:

$$\delta_{ii} = 3, \quad \delta_{ij}u_i = u_j, \quad \delta_{ij}\delta_{jk} = \delta_{ik}. \tag{1.10}$$

Another commonly used symbol is the permutation symbol ε_{ijk}, which is defined as follows:

$$\varepsilon_{ijk} = \begin{cases} 1, & \text{for even permutations of } (i,j,k) \text{ (i.e. } 123, 231, 312). \\ -1, & \text{for odd permutations of } (i,j,k) \text{ (i.e. } 132, 213, 321). \\ 0, & \text{if there is a repeated index.} \end{cases} \tag{1.11}$$

The permutation symbol has the following properties:

$$\varepsilon_{ijk} = \varepsilon_{jki} = \varepsilon_{kij}, \quad \varepsilon_{ijk} = -\varepsilon_{ikj}, \quad \varepsilon_{ijk} = -\varepsilon_{jik}. \tag{1.12}$$

1.1.3 Tensor

A second-order tensor can be viewed as a linear operator acting on a vector $\mathbf{u}$, producing a new vector $\mathbf{v}$ as

$$\mathbf{v} = \mathbf{A}\mathbf{u}. \tag{1.13}$$

A second-order tensor can be written in the index notation as

$$\mathbf{A} = A_{ij}\mathbf{e}_i \otimes \mathbf{e}_j. \tag{1.14}$$

where $\otimes$ is the dyad of vectors.

A unit tensor $\mathbf{I}$ can be represented using the Kronecker delta symbol as

$$\mathbf{I} = \delta_{ij}\mathbf{e}_i \otimes \mathbf{e}_j = \mathbf{e}_i \otimes \mathbf{e}_i. \tag{1.15}$$

The dot product of two second-order tensors $\mathbf{A}$ and $\mathbf{B}$ is typically denoted as $\mathbf{AB}$, which is still a second-order tensor. The components of the dot product $\mathbf{AB}$ can be represented as

$$(\mathbf{AB})_{ij} = A_{ik}B_{kj}. \tag{1.16}$$

The double dot product of two second-order tensors results in a scalar, as given by the following expression:

$$\mathbf{A} : \mathbf{B} = A_{ij}\mathbf{e}_i \otimes \mathbf{e}_j : B_{kl}\mathbf{e}_k \otimes \mathbf{e}_l = A_{ij}B_{kl}(\mathbf{e}_i \cdot \mathbf{e}_k)(\mathbf{e}_j \cdot \mathbf{e}_l) = A_{ij}B_{ij}. \tag{1.17}$$

The transpose of a second-order tensor $\mathbf{A}$ is denoted as $\mathbf{A}^{\mathrm{T}}$, with the following property:

$$(A^{\mathrm{T}})_{ij} = A_{ji}. \tag{1.18}$$

The trace of a second-order tensor $\mathbf{A}$ equals to the sum of the diagonal terms represented as

$$\mathrm{tr}\,(\mathbf{A}) = A_{ii}. \tag{1.19}$$

The determinant of a tensor $\mathbf{A}$ is the determinant of the matrix $\mathbf{A}$, represented as

$$\det(\mathbf{A}) = \varepsilon_{ijk} A_{1i} A_{2j} A_{3k}. \tag{1.20}$$

For a nonsingular tensor, namely, $\det(\mathbf{A}) \neq 0$, there exists a unique inverse of the tensor, denoted as $\mathbf{A}^{-1}$, which satisfies as

$$\mathbf{A}\mathbf{A}^{-1} = \mathbf{I}. \tag{1.21}$$

A tensor can be decomposed into a symmetric part and a skew part as

$$\mathrm{sym}(\mathbf{A}) = \frac{\mathbf{A} + \mathbf{A}^{\mathrm{T}}}{2}, \quad \mathrm{skew}(\mathbf{A}) = \frac{\mathbf{A} - \mathbf{A}^{\mathrm{T}}}{2}. \tag{1.22}$$

A second-order tensor has different component forms in different coordinate systems:

$$\mathbf{A} = A_{ij} \mathbf{e}_i \otimes \mathbf{e}_j \quad \text{or} \quad \mathbf{A} = \tilde{A}_{ij} \tilde{\mathbf{e}}_i \otimes \tilde{\mathbf{e}}_j. \tag{1.23}$$

These two component forms can be related through the transformation of the basis vectors corresponding to the two coordinate systems, as

$$\mathbf{e}_i = Q_{ij} \tilde{\mathbf{e}}_j \quad Q_{ij} = \mathbf{e}_i \cdot \tilde{\mathbf{e}}_j. \tag{1.24}$$

where Q_{ij} represents the direction cosines between the basis vectors of the two coordinate systems. $\mathbf{Q}$ is an orthogonal second-order matrix with $\mathbf{Q}^{\mathrm{T}} = \mathbf{Q}^{-1}$.

Furthermore, we have:

$$\tilde{A}_{ij} = Q_{ki} A_{kl} Q_{lj}, \quad [\tilde{\mathbf{A}}] = [\mathbf{Q}]^{\mathrm{T}} [\mathbf{A}][\mathbf{Q}]. \tag{1.25}$$

A vector can be treated as a tensor with order 1. Thus, higher order tensors can also be defined in a similar way. For example, tensors with an order 3 or 4 can be defined as

$$\mathcal{A} = A_{ijk} \mathbf{e}_i \otimes \mathbf{e}_j \otimes \mathbf{e}_k. \tag{1.26a}$$

$$\mathbb{A} = A_{ijkl} \mathbf{e}_i \otimes \mathbf{e}_j \otimes \mathbf{e}_k \otimes \mathbf{e}_l. \tag{1.26b}$$

1.1.4 Gradient, Divergence, and Curl

We introduce the vector operator ∇, commonly referred to as the Nabla operator, which is frequently used in computations:

$$\nabla(\bullet) = \frac{\partial(\bullet)}{\partial x_i} \mathbf{e}_i = \frac{\partial(\bullet)}{\partial x_1} \mathbf{e}_1 + \frac{\partial(\bullet)}{\partial x_2} \mathbf{e}_2 + \frac{\partial(\bullet)}{\partial x_3} \mathbf{e}_3. \tag{1.27}$$

Thus, the gradient vector of a scalar field at any point is given as

$$\mathrm{grad}\Phi = \nabla\Phi = \frac{\partial \Phi}{\partial x_i}\mathbf{e}_i. \tag{1.28}$$

The gradient of a vector is a second-order tensor, defined as follows:

$$\mathrm{grad}\mathbf{u} = \nabla\mathbf{u} = \frac{\partial u_i}{\partial x_j}\mathbf{e}_i \otimes \mathbf{e}_j. \tag{1.29}$$

Similarly, the gradient operator can also be applied to a tensor. For example, the gradient of a second-order tensor **A** can be written as

$$\mathrm{grad}\mathbf{A} = \nabla\mathbf{A} = \frac{\partial A_{ij}}{\partial x_k}\mathbf{e}_i \otimes \mathbf{e}_j \otimes \mathbf{e}_k. \tag{1.30}$$

The Nabla operator can also be combined with a vector through dot and cross products. The dot product of the Nabla operator with a vector is referred to as the divergence of the vector:

$$\mathrm{div}\mathbf{u} = \nabla \cdot \mathbf{u} = \frac{\partial u_j}{\partial x_i}\mathbf{e}_j \cdot \mathbf{e}_i = \frac{\partial u_j}{\partial x_i}\delta_{ij} = \frac{\partial u_i}{\partial x_i}. \tag{1.31}$$

The divergence of a second-order tensor is represented as

$$\mathrm{div}\mathbf{A} = \nabla \cdot \mathbf{A} = \frac{\partial A_{ij}}{\partial x_k}\mathbf{e}_i \otimes \mathbf{e}_j \cdot \mathbf{e}_k = \frac{\partial A_{ij}}{\partial x_k}\mathbf{e}_i\delta_{jk} = \frac{\partial A_{ij}}{\partial x_j}\mathbf{e}_i. \tag{1.32}$$

If the Nabla operator is dotted with the gradient of a scalar field, the result is

$$\nabla \cdot \nabla\Phi = (\nabla \cdot \nabla)\Phi = \nabla^2\Phi = \frac{\partial^2 \Phi}{\partial x_1^2} + \frac{\partial^2 \Phi}{\partial x_2^2} + \frac{\partial^2 \Phi}{\partial x_3^2}. \tag{1.33}$$

The ∇^2 operator is known as the Laplacian operator.

The cross product of the Nabla operator with a vector is referred to as the curl of the vector:

$$\mathrm{curl}\mathbf{u} = \nabla \times \mathbf{u} = \frac{\partial u_j}{\partial x_i}\mathbf{e}_i \times \mathbf{e}_j = \varepsilon_{ijk}\frac{\partial u_j}{\partial x_i}\mathbf{e}_k, \tag{1.34}$$

which indicates the curl of a vector field representing the degree of rotation of the field vectors around any point.

1.2 Kinematics

1.2.1 Deformation Gradient

Different from the small strain condition, finite deformation needs to consider higher order geometric and mechanical quantities. The analysis based on the undeformed or deformed configuration leads to different descriptions of the same deformation: material (Lagrangian) description or spatial (Eulerian) description.

In the theory of continuum mechanics, a body is assumed to possess continuity in both space and time (or at least be piecewise continuous), allowing its various physical quantities to be represented by continuous functions. As illustrated in Figure 1.1, the continuous

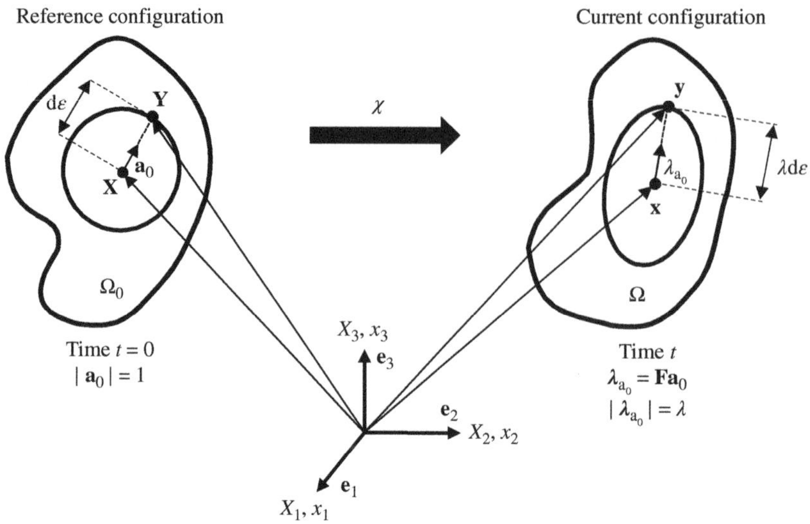

Figure 1.1 Deformation of a continuous medium.

spatial regions it occupies before and after a time interval are denoted by Ω_0 and Ω, respectively. The position of any point between these two configurations is related by a mapping:

$$\mathbf{x} = \chi(\mathbf{X}, t), \quad \mathbf{X} = \chi^{-1}(\mathbf{x}, t), \tag{1.35}$$

where $\mathbf{x}$ represents the position of any point in the current configuration and $\mathbf{X}$ represents the corresponding position of that point in the reference configuration.

With the coordinates of each point in both the reference configuration and the current configuration, the displacement of each point can be computed as follows:

$$\mathbf{u} = \mathbf{x} - \mathbf{X}. \tag{1.36}$$

The deformation gradient $\mathbf{F}$ is then introduced to relate the point vector in the reference configuration and that in the current configuration, which is represented as

$$\mathbf{F} = \frac{\partial \mathbf{x}}{\partial \mathbf{X}}, \quad F_{aA} = \frac{\partial x_a}{\partial X_A}. \tag{1.37}$$

We can further derive the relationship between the deformation gradient tensor and the displacement field as follows:

$$\mathbf{F} = \mathbf{I} + \frac{\partial \mathbf{u}}{\partial \mathbf{X}}, \quad F_{aA} = \delta_{aA} + \frac{\partial u_a}{\partial X_A}. \tag{1.38}$$

We can then easily relate the vector line in the reference and current configurations as

$$\mathrm{d}\mathbf{x} = \mathbf{F}\mathrm{d}\mathbf{X}. \tag{1.39}$$

The determinant of the deformation gradient matrix det$\mathbf{F}$ (or J) represents the deformation state of the infinitesimal volume element as

$$\mathrm{d}v = J\mathrm{d}V, \tag{1.40}$$

where $\mathrm{d}V$ and $\mathrm{d}v$ are the infinitesimal volume elements defined in the undeformed and deformed configurations, respectively. When $\det\mathbf{F} = 1$, the deformation is isochoric, meaning the infinitesimal volume element remains constant.

The infinitesimal volume element can be calculated through the dot product of a small surface and a material line element as

$$\mathrm{d}V = \mathrm{d}\mathbf{S} \cdot \mathrm{d}\mathbf{X}, \tag{1.41a}$$

$$\mathrm{d}v = \mathrm{d}\mathbf{s} \cdot \mathrm{d}\mathbf{x}, \tag{1.41b}$$

where $\mathrm{d}\mathbf{S}$ and $\mathrm{d}\mathbf{s}$ are the infinitesimally small areas in the reference and current configurations, respectively, and $\mathrm{d}\mathbf{X}$ and $\mathrm{d}\mathbf{x}$ are the line elements in the reference and current configurations, respectively.

Considering Eqs. (1.39)–(1.41), we can obtain:

$$\mathrm{d}\mathbf{s} = J\mathbf{F}^{-\mathrm{T}}\mathrm{d}\mathbf{S}. \tag{1.42}$$

This equation is known as Nanson's formula.

The deformation gradient contains both the contribution of rotation and stretch. To decompose the two effects, the deformation gradient $\mathbf{F}$ can be decomposed into

$$\mathbf{F} = \mathbf{R}\mathbf{U} = \mathbf{v}\mathbf{R}, \tag{1.43}$$

where $\mathbf{R}$ is an orthogonal rotation tensor, and $\mathbf{U}$ and $\mathbf{v}$ correspond to the right and left stretch tensors, respectively.

The time derivative of the deformation gradient $\mathbf{F}$ can be represented as

$$\dot{\mathbf{F}} = \frac{\partial}{\partial t}\left(\frac{\partial \mathbf{x}(\mathbf{X}, t)}{\partial \mathbf{X}}\right) = \frac{\partial}{\partial \mathbf{X}}\left(\frac{\partial \mathbf{x}(\mathbf{X}, t)}{\partial t}\right) = \frac{\partial \mathbf{V}}{\partial \mathbf{X}} = \mathrm{Grad}\mathbf{V}, \tag{1.44}$$

where $\mathbf{V}$ is the velocity defined in the reference configuration, and $\mathrm{Grad}(\cdot)$ is the gradient operator defined in the reference configuration.

Note: Different from the small strain condition, the operations, such as the gradient and divergence, need to be specified in the reference or current configurations. The relationship between these operations can be built using the chain rule [1].

The spatial velocity gradient can be further defined as

$$\mathbf{L} = \frac{\partial \mathbf{v}}{\partial \mathbf{x}} = \frac{\partial \mathbf{v}}{\partial \mathbf{X}}\frac{\partial \mathbf{X}}{\partial \mathbf{x}} = \frac{\partial}{\partial t}\left(\frac{\partial \mathbf{x}(\mathbf{X}, t)}{\partial \mathbf{X}}\right)\frac{\partial \mathbf{X}}{\partial \mathbf{x}} = \dot{\mathbf{F}}\mathbf{F}^{-1}. \tag{1.45}$$

1.2.2 Strain Tensor

We further measure the change in the squared distance between any material point and its neighboring point, represented as

$$\begin{aligned}
&(\mathrm{d}l)^2 - (\mathrm{d}L)^2 = \mathrm{d}\mathbf{x} \cdot \mathrm{d}\mathbf{x} - \mathrm{d}\mathbf{X} \cdot \mathrm{d}\mathbf{X} \\
&= \mathrm{d}\mathbf{X} \cdot \mathbf{F}^{\mathrm{T}} \cdot \mathbf{F} \cdot \mathrm{d}\mathbf{X} - \mathrm{d}\mathbf{X} \cdot \mathrm{d}\mathbf{X} \\
&= \mathrm{d}\mathbf{X} \cdot (\mathbf{F}^{\mathrm{T}} \cdot \mathbf{F} - \mathbf{I}) \cdot \mathrm{d}\mathbf{X},
\end{aligned} \tag{1.46}$$

where $\mathrm{d}L$ and $\mathrm{d}l$ are the length in the reference and deformed configurations, respectively.

The previous equation provides the derivation of the Green–Lagrange strain tensor, which can be expressed as

$$\mathbf{E} = \frac{1}{2}(\mathbf{F}^{\mathrm{T}} \cdot \mathbf{F} - \mathbf{I}), \tag{1.47a}$$

$$E_{AB} = \frac{1}{2}(F_{aA}F_{aB} - \delta_{AB}). \tag{1.47b}$$

The square of the right stretch tensor and the left stretch tensor are, respectively, known as the right Cauchy–Green tensor **C** and left Cauchy–Green tensor **b**:

$$\mathbf{C} = \mathbf{U}^2 = \mathbf{F}^{\mathrm{T}}\mathbf{F}, \quad \mathbf{b} = \mathbf{v}^2 = \mathbf{F}\mathbf{F}^{\mathrm{T}}, \tag{1.48a}$$

$$C_{AB} = F_{aA}F_{aB}, \quad b_{ab} = F_{aA}F_{bA}. \tag{1.48b}$$

Another common strain tensor is Euler–Almansi strain tensor **e**, defined as

$$\mathbf{e} = \frac{1}{2}(\mathbf{I} - \mathbf{F}^{-\mathrm{T}} \cdot \mathbf{F}^{-1}), \tag{1.49a}$$

$$e_{ab} = \frac{1}{2}(\delta_{ab} - F^{-1}_{Ca}F^{-1}_{Cb}). \tag{1.49b}$$

The right and left Cauchy–Green tensors share the same eigenvalues. The three invariants I_1, I_2, I_3 can be represented as

$$I_1 = C_{ii}, \tag{1.50a}$$

$$I_2 = \frac{1}{2}(C_{ii}C_{jj} - C_{ij}C_{ij}), \tag{1.50b}$$

$$I_3 = \det(\mathbf{C}). \tag{1.50c}$$

If the three principal stretches of the deformation gradient **F** are denoted as λ_1, λ_2, and λ_3, the three invariants can then be represented as

$$I_1 = \lambda_1^2 + \lambda_2^2 + \lambda_3^2, \tag{1.51a}$$

$$I_2 = \lambda_1^2\lambda_2^2 + \lambda_2^2\lambda_3^2 + \lambda_1^2\lambda_3^2, \tag{1.51b}$$

$$I_3 = \lambda_1^2\lambda_2^2\lambda_3^2. \tag{1.51c}$$

1.3 Stress

The force exerted per unit area on an object is referred to as stress. To evaluate the stress state at a specific point within an object, we can hypothetically cut the object with a plane passing through that point, allowing us to obtain the stress at the point on the plane, which includes both magnitude and direction. Since there are infinitely many planes that can pass through a given point, it is impractical to describe the stress state at that point using an infinite number of stress vectors. This necessitates the introduction of the second-order stress tensor. The stress tensor effectively includes the information of the magnitude of the stress at a point, along with two directional components: one corresponding to the normal of the cutting plane and the other representing the projection direction of the stress tensor on that plane.

For an arbitrary plane, the force per unit area acting on this plane in the normal direction $\mathbf{n}$ is denoted as the Cauchy traction vector $\mathbf{t}$. The Cauchy traction vector is then related to the Cauchy stress as

$$\mathbf{t} = \boldsymbol{\sigma}\mathbf{n}. \tag{1.52}$$

The Cauchy stress $\boldsymbol{\sigma}$ is a second-order symmetric stress tensor defined in the current configuration and is also referred to as the true stress tensor.

Correspondingly, we can also define the corresponding traction vector $\mathbf{T}$ in the reference configuration. The first Piola–Kirchhoff stress tensor $\mathbf{P}$ is related to the traction $\mathbf{T}$ as

$$\mathbf{T} = \mathbf{P}\mathbf{N}, \tag{1.53}$$

where $\mathbf{N}$ is the normal direction of the plane in the reference configuration.

The total force applied on the plane should be the same regardless of the configurations, which can be represented as

$$\mathbf{t}\mathrm{d}s = \mathbf{T}\mathrm{d}S, \tag{1.54}$$

where $\mathrm{d}s$ and $\mathrm{d}S$ are the areas in the current configuration and the reference configuration, respectively.

Combining the preceding three equations yields

$$\boldsymbol{\sigma}(\mathbf{x}, t)\mathbf{n}\mathrm{d}s = \mathbf{P}(\mathbf{X}, t)\mathbf{N}\mathrm{d}S, \tag{1.55a}$$

$$\boldsymbol{\sigma}(\mathbf{x}, t)\mathrm{d}\mathbf{s} = \mathbf{P}(\mathbf{X}, t)\mathrm{d}\mathbf{S}. \tag{1.55b}$$

Using Nanson's equation in Eq. (1.42), we can obtain the relationship between the first Piola–Kirchhoff stress and the Cauchy stress as

$$\mathbf{P} = J\boldsymbol{\sigma}\mathbf{F}^{-\mathrm{T}}. \tag{1.56}$$

Since the deformation gradient tensor $\mathbf{F}$ is not a symmetric tensor in general, $\mathbf{P}$ is also an asymmetric tensor in most cases. The second Piola–Kirchhoff stress tensor $\mathbf{S}$, as a symmetric tensor, is further defined as

$$\mathbf{S} = \mathbf{F}^{-1}\mathbf{P} = J\mathbf{F}^{-1}\boldsymbol{\sigma}\mathbf{F}^{-\mathrm{T}}. \tag{1.57}$$

1.4 Balance Principles

Classic balance principles, such as conservation of mass, balance of momentum, and balance of energy, should also be satisfied in continuum mechanics. By applying the second law of thermodynamics to continuum mechanics, it can be expressed as an inequality governing the rate of change of free energy, thus imposing mathematical constraints on material deformation. In this chapter, we will introduce these balance principles together with the entropy inequality. It should be noted that the presentation in this book omits many details. For a complete derivative process, the readers can refer to classic continuum mechanics monographs [1–4].

1.4.1 Material Derivative and Spatial Derivative

The spatial derivative is defined as the time derivative of a physical or geometric quantity, with the current position **x** fixed, which can be expressed as

$$\frac{\partial \Phi}{\partial t} = \frac{\partial}{\partial t}\Phi(\mathbf{x}, t)\Big|_{\mathbf{x}}. \tag{1.58}$$

The material derivative is defined as the time derivative of a physical or geometric quantity, with the referential position **X** fixed, which can be expressed as:

$$\frac{\mathrm{d}\Phi}{\mathrm{d}t} = \frac{\partial}{\partial t}\Phi(\mathbf{x}, t)\Big|_{\mathbf{x}} + \frac{\partial \Phi}{\partial \mathbf{x}}\frac{\partial \mathbf{x}}{\partial t} = \frac{\partial}{\partial t}\Phi(\mathbf{x}, t)\Big|_{\mathbf{x}} + \mathrm{grad}\Phi \cdot \mathbf{v}. \tag{1.59}$$

1.4.2 Reynolds Transport Theorem

Suppose a body occupies a spatial domain Ω in three-dimensional space, with a volume v. Inside this domain, there exists a scalar field $\Phi = \Phi(\mathbf{x}, t)$. The time derivative of the integral of Φ over this domain cannot be directly evaluated by moving the derivative operation inside the integration function due to a continuous change of the volume element v. To compute this time derivative, we can use the following procedures:

$$\begin{aligned}
\frac{\mathrm{d}}{\mathrm{d}t}\int_{\Omega} \Phi(\mathbf{x}, t)\mathrm{d}v &= \frac{\mathrm{d}}{\mathrm{d}t}\int_{\Omega} \Phi(\mathbf{x}(\mathbf{X}, t), t)J\mathrm{d}V \\
&= \int_{\Omega_0} \left(J\frac{\mathrm{d}\Phi(\mathbf{x}(\mathbf{X}, t), t)}{\mathrm{d}t} + \Phi(\mathbf{x}(\mathbf{X}, t), t)\dot{J}\right)\mathrm{d}V \\
&= \int_{\Omega} \left(\frac{\mathrm{d}\Phi(\mathbf{x}(\mathbf{X}, t), t)}{\mathrm{d}t} + \Phi(\mathbf{x}(\mathbf{X}, t), t)\mathrm{div}\mathbf{v}\right)\mathrm{d}v.
\end{aligned} \tag{1.60}$$

In this derivative, we have used the relationship that $\dot{J}/J = \mathrm{div}\mathbf{v}$. This function is known as Reynolds Transport Theorem.

Using the relation in Eq. (1.59), Eq. (1.60) can be further rewritten as

$$\begin{aligned}
\frac{\mathrm{d}}{\mathrm{d}t}\int_{\Omega} \Phi(\mathbf{x}, t)\mathrm{d}v &= \int_{\Omega} \left(\frac{\partial \Phi}{\partial t} + \mathrm{grad}\Phi \cdot \mathbf{v} + \Phi\mathrm{div}\mathbf{v}\right)\mathrm{d}v \\
&= \int_{\Omega} \left(\frac{\partial \Phi}{\partial t} + \mathrm{div}(\Phi\mathbf{v})\right)\mathrm{d}v \\
&= \int_{\Omega} \frac{\partial \Phi}{\partial t}\mathrm{d}v + \int_{\partial\Omega} \Phi\mathbf{v} \cdot \mathbf{n}\mathrm{d}s.
\end{aligned} \tag{1.61}$$

This equation is another form of Reynolds Transport Theorem. It shows the rate of change of the field Φ has two sources. The first part arises from the local rate of the field Φ within the region Ω. The second part is caused by the moving region.

1.4.3 Conservation of Mass

In a closed system, the total mass of material remains constant throughout motion and deformation for any chosen configuration. Mathematically, this can be expressed as the equivalence of integral of mass density over the volume in reference configuration and integral of mass density over the volume in current configuration:

$$\int_{\Omega_0} \rho_0(\mathbf{X})\mathrm{d}V = \int_{\Omega} \rho(\mathbf{x})\mathrm{d}v, \tag{1.62}$$

where ρ_0 is the mass density in reference configuration Ω_0 and ρ is the mass density in current configuration Ω. When this conclusion holds for an arbitrarily small neighborhood around each material point, it implies that the mass is locally conserved. This local conservation requirement can be expressed as

$$\rho_0 = J\rho. \tag{1.63}$$

Since $\dot{\rho}_0 = 0$, this gives $\dot{\overline{(J\rho)}} = 0$. This equation can be further rewritten as

$$\dot{\rho} + \rho \mathrm{div}\mathbf{v} = 0, \tag{1.64a}$$

$$\frac{\partial \rho}{\partial t} + \nabla \cdot (\rho \mathbf{v}) = 0. \tag{1.64b}$$

1.4.4 Balance of Momentum

Suppose a body with mass density ρ occupies a region Ω in its current configuration, with boundary $\partial\Omega$. The external force acting on the body can be divided into two components: one part is the body force $\mathbf{b}$ per unit volume within the region Ω, and the other part is the surface force $\mathbf{t}$ per unit area on the boundary $\partial\Omega$. The change rate of the momentum should be equal to the total applied force, represented as

$$\frac{\mathrm{d}}{\mathrm{d}t}\int_{\Omega} \rho\mathbf{v}\mathrm{d}v = \int_{\Omega} \mathbf{b}\mathrm{d}v + \oint_{\partial\Omega} \mathbf{t}\mathrm{d}s. \tag{1.65}$$

The surface integral of force can be expressed as

$$\oint_{\partial\Omega} \mathbf{t}\mathrm{d}s = \oint_{\partial\Omega} \boldsymbol{\sigma}\mathbf{n}\mathrm{d}s = \int_{\Omega} \nabla \cdot \boldsymbol{\sigma}\mathrm{d}v. \tag{1.66}$$

Using Reynolds Transport Theorem and the mass conservation law, we can then obtain the following expression [1]:

$$\frac{\mathrm{d}}{\mathrm{d}t}\int_{\Omega} \rho\mathbf{v}\mathrm{d}v = \int_{\Omega} \rho\frac{\mathrm{d}\mathbf{v}}{\mathrm{d}t}\mathrm{d}v. \tag{1.67}$$

The law of conservation of momentum can then be written as

$$\int_{\Omega} (\nabla \cdot \boldsymbol{\sigma} + \mathbf{b})\mathrm{d}v = \int_{\Omega} \rho \frac{\mathrm{d}\mathbf{v}}{\mathrm{d}t}\mathrm{d}v. \tag{1.68}$$

Since the size of the region Ω is arbitrary, the previous expression can be written as the local form of the momentum conservation equation in the current configuration:

$$\nabla \cdot \boldsymbol{\sigma} + \mathbf{b} = \rho \frac{\mathrm{d}\mathbf{v}}{\mathrm{d}t}. \tag{1.69}$$

Under the condition of neglecting inertial term, the preceding equation degenerates to the equilibrium equation in elasticity:

$$\nabla \cdot \boldsymbol{\sigma} + \mathbf{b} = \mathbf{0}. \tag{1.70}$$

Using a similar process, the local form of the momentum conservation equation in the reference configuration can be obtained as

$$\nabla_{\mathrm{X}} \cdot \mathbf{P} + \mathbf{B} = \rho_0 \frac{\mathrm{d}\mathbf{V}}{\mathrm{d}t}, \tag{1.71}$$

where $\mathbf{B}$ is the body force in reference configuration.

1.4.5 Balance of Angular Momentum

The conservation of angular momentum states that the rate of change of the angular momentum about a certain point is equal to the resultant moment of external forces at that point. Consider a body with mass density ρ occupies a region Ω in its current configuration, with boundary $\partial\Omega$. The external moment can be divided into two components: one part is caused by the body force per unit volume within the region Ω, and the other part is induced by the surface force per unit area on the boundary $\partial\Omega$. Let $\mathbf{r}$ be the vector from the coordinate origin to a point on the body. Therefore, the balance of angular momentum in the current configuration can be expressed as

$$\frac{\mathrm{d}}{\mathrm{d}t}\int_{\Omega} \mathbf{r} \times \rho\mathbf{v}\mathrm{d}v = \int_{\Omega} \mathbf{r} \times \mathbf{b}\mathrm{d}v + \oint_{\partial\Omega} \mathbf{r} \times \mathbf{t}\mathrm{d}s. \tag{1.72}$$

According to Reynolds Transport Theorem, the left-hand side of the equation can be simplified as

$$\frac{\mathrm{d}}{\mathrm{d}t}\int_{\Omega} \mathbf{r} \times \rho\mathbf{v}\mathrm{d}v = \int_{\Omega} \rho\mathbf{r} \times \frac{\mathrm{d}\mathbf{v}}{\mathrm{d}t}\mathrm{d}v. \tag{1.73}$$

The moment of the resultant external forces can be simplified to:

$$\int_{\Omega} \mathbf{r} \times \mathbf{b}\mathrm{d}v + \oint_{\partial\Omega} \mathbf{r} \times \mathbf{t}\mathrm{d}s = \int_{\Omega} [\mathbf{r} \times (\nabla \cdot \boldsymbol{\sigma} + \mathbf{b}) + \mathcal{E} : \boldsymbol{\sigma}^{\mathrm{T}}]\mathrm{d}v, \tag{1.74}$$

where $\mathcal{E}$ is the third-order permutation tensor.

Then, the law of conservation of angular momentum can be written as

$$\int_{\Omega} \left[\mathbf{r} \times \left(\nabla \cdot \boldsymbol{\sigma} + \mathbf{b} - \rho \frac{\mathrm{d}\mathbf{v}}{\mathrm{d}t} \right) + \mathcal{E} : \boldsymbol{\sigma}^{\mathrm{T}} \right] \mathrm{d}v = 0. \tag{1.75}$$

This equation holds for regions of arbitrary size. Combining it with the balance of momentum, we obtain the following:

$$\mathcal{E} : \boldsymbol{\sigma}^{\mathrm{T}} = 0, \tag{1.76}$$

$$\sigma_{ij} = \sigma_{ji}, \quad \boldsymbol{\sigma} = \boldsymbol{\sigma}^{\mathrm{T}}. \tag{1.77}$$

Thus, the balance of the angular momentum requires that the Cauchy stress tensor is a symmetric second-order tensor.

1.4.6 Balance of Mechanical Energy

The total energy of an object consists of kinetic energy and internal energy, while we consider only mechanical energy and neglect other forms of energy such as thermal, electrical, or chemical energy. The balance of mechanical energy can be stated as follows: the material time derivative of the total energy equals to the power of the external forces acting on the volume domain. This can be expressed as

$$\dot{K} + P_{\mathrm{int}} = P_{\mathrm{ext}}, \tag{1.78}$$

where K is the kinetic energy, P_{int} is the rate of internal mechanical work, and P_{ext} is the rate of external mechanical work.

The kinetic energy can be represented as

$$K = \int_{\Omega} \frac{1}{2} \rho \mathbf{v} \cdot \mathbf{v} \mathrm{d}v. \tag{1.79}$$

Using Reynolds Transport Theorem, the rate of the kinetic energy can be written as

$$\dot{K} = \frac{\mathrm{d}}{\mathrm{d}t} \int_{\Omega} \frac{1}{2} \rho \mathbf{v} \cdot \mathbf{v} \mathrm{d}v = \int_{\Omega} \rho \mathbf{v} \cdot \frac{\mathrm{d}\mathbf{v}}{\mathrm{d}t} \mathrm{d}v. \tag{1.80}$$

The external mechanical work can be written as

$$\begin{aligned} P_{\mathrm{ext}} &= \int_{\Omega} \mathbf{b} \cdot \mathbf{v} \mathrm{d}v + \oint_{\partial\Omega} \mathbf{t} \cdot \mathbf{v} \mathrm{d}s \\ &= \int_{\Omega} \mathbf{b} \cdot \mathbf{v} \mathrm{d}v + \oint_{\partial\Omega} \boldsymbol{\sigma}\mathbf{n} \cdot \mathbf{v} \mathrm{d}s \\ &= \int_{\Omega} \mathbf{b} \cdot \mathbf{v} \mathrm{d}v + \int_{\Omega} \mathrm{div}(\boldsymbol{\sigma}^{\mathrm{T}}\mathbf{v}) \mathrm{d}v \\ &= \int_{\Omega} \mathbf{b} \cdot \mathbf{v} \mathrm{d}v + \int_{\Omega} (\mathrm{div}\boldsymbol{\sigma} \cdot \mathbf{v} + \boldsymbol{\sigma} : \mathrm{grad}\mathbf{v}) \mathrm{d}v. \end{aligned} \tag{1.81}$$

Combining the preceding four equations together with the balance of momentum (Eq. 1.69), it can be obtained that

$$P_{\text{int}} = \int_{\Omega} \boldsymbol{\sigma} : \text{grad}\mathbf{v}\text{d}v = \int_{\Omega} \boldsymbol{\sigma} : \mathbf{L}\text{d}v = \int_{\Omega} \boldsymbol{\sigma} : \mathbf{d}\text{d}v, \tag{1.82}$$

where $\mathbf{d}$ is the symmetric part of the spatial velocity gradient tensor.

Note. Various stress and strain tensors have been defined in the previous part. Stress and strain tensors defined within the same description method are thermodynamic conjugate variables. The rate of change in the deformation energy density can also be expressed using other stress and strain pairs using the relation: $J\boldsymbol{\sigma} : \mathbf{d} = \mathbf{P} : \dot{\mathbf{F}} = \mathbf{S} : \dot{\mathbf{E}}$.

1.4.7 Balance of Energy

For a system with both mechanical and thermal energy, the first law of thermodynamics should be stated as the increase in the kinetic energy and internal energy should be equal to the external mechanical power and thermal power, represented as

$$\frac{\text{d}}{\text{d}t}K + \frac{\text{d}}{\text{d}t}\int_{\Omega} e_{\text{c}}\text{d}V = P_{\text{ext}} + Q(t), \tag{1.83}$$

where e_{c} is the internal energy and $Q(t)$ is the rate of thermal work in the current configuration.

The kinetic energy and external mechanical power are the same as Eqs. (1.79) and (1.81). The thermal power can be defined as

$$\begin{aligned} Q(t) &= -\oint_{\partial\Omega} \mathbf{q} \cdot \mathbf{n}\text{d}s + \int_{\Omega} r\text{d}v \\ &= -\int_{\Omega} \text{div}\mathbf{q}\text{d}v + \int_{\Omega} r\text{d}v, \end{aligned} \tag{1.84}$$

where $\mathbf{q}$ is the heat flux through the surface, namely, heat per unit time and per unit area, and r is the heat source per unit time and per unit volume.

Using the same procedure, we can obtain the following form of the first law in current configuration as

$$\frac{\text{d}}{\text{d}t}\int_{\Omega} e_{\text{c}}\text{d}v = \int_{\Omega} (\boldsymbol{\sigma} : \mathbf{d} - \text{div}\mathbf{q} + r)\text{d}v. \tag{1.85}$$

The first law of thermodynamics can also be rewritten in the material configuration

$$\frac{\text{d}}{\text{d}t}\int_{\Omega_0} \left(\frac{1}{2}\rho_0 \mathbf{V} \cdot \mathbf{V} + e\right)\text{d}V = \int_{\Omega_0} (\mathbf{B} \cdot \mathbf{V} + R)\text{d}V + \oint_{\partial\Omega_0} (\mathbf{T} \cdot \mathbf{V} - \mathbf{Q} \cdot \mathbf{N})\text{d}S, \tag{1.86}$$

where e is the internal energy in reference configuration, R is the heat source in reference configuration, and $\mathbf{Q}$ is the heat flux in reference configuration.

Using the same procedure, the previous equation can be further reduced to the following form:

$$\frac{\mathrm{d}}{\mathrm{d}t}\int_{\Omega_0} e\mathrm{d}V = \int_{\Omega_0} (\mathbf{P} : \dot{\mathbf{F}} - \mathrm{Div}\mathbf{Q} + R)\mathrm{d}V. \tag{1.87}$$

Different from the first law in current configuration (Eq. 1.85), Eq. (1.87) can be written in the local form as

$$\dot{e} = \mathbf{P} : \dot{\mathbf{F}} - \mathrm{Div}\mathbf{Q} + R. \tag{1.88}$$

1.4.8 Entropy Inequality

We will then introduce the second law of thermodynamics, which governs the direction of energy transfer. The second law requires that in any process, the system evolves in a direction towards increasing entropy η. We first define the entropy flux $\mathbf{h}$ and the entropy source $\tilde{r}$ in the current configuration as

$$\mathbf{h} = \frac{\mathbf{q}}{T}, \quad \tilde{r} = \frac{r}{T}, \tag{1.89}$$

where T is the temperature.

The total entropy production of the whole system should not be negative, which gives

$$\frac{\mathrm{d}}{\mathrm{d}t}\int_{\Omega} \eta_c \mathrm{d}v + \oint_{\partial\Omega} \frac{\mathbf{q}}{T}\cdot\mathbf{n}\mathrm{d}s - \int_{\Omega} \frac{r}{T}\mathrm{d}v \geq 0, \tag{1.90}$$

where η_c is the entropy in the current configuration. This is the entropy inequality in the current configuration. The previous form cannot be directly written in the local form.

In order to obtain the local form of the entropy inequality, we need to write the entropy inequality in the reference configuration. Similarly, the entropy flux $\mathbf{H}$ and the entropy source $\widetilde{R}$ in the reference configuration can be represented as

$$\mathbf{H} = \frac{\mathbf{Q}}{T}, \quad \widetilde{R} = \frac{R}{T}. \tag{1.91}$$

The second law is then written as

$$\frac{\mathrm{d}}{\mathrm{d}t}\int_{\Omega_0} \eta \mathrm{d}V + \oint_{\partial\Omega_0} \frac{\mathbf{Q}}{T}\cdot\mathbf{N}\mathrm{d}S - \int_{\Omega_0} \frac{R}{T}\mathrm{d}V \geq 0, \tag{1.92}$$

where η is entropy in the reference configuration.

Using the integral theorem, the equation can be rewritten as

$$\int_{\Omega_0} \left(\dot{\eta} + \frac{\mathrm{Div}\mathbf{Q}}{T} - \frac{\mathbf{Q}\cdot\mathrm{Grad}T}{T^2} - \frac{R}{T}\right)\mathrm{d}V \geq 0. \tag{1.93}$$

The local form of the second law of thermodynamics can then be written as

$$\dot{\eta} + \frac{\mathrm{Div}\mathbf{Q}}{T} - \frac{\mathbf{Q}\cdot\mathrm{Grad}T}{T^2} - \frac{R}{T} \geq 0. \tag{1.94}$$

Through elimination of the heat source R using Eq. (1.88), the preceding equation can be represented as

$$\mathbf{P} : \dot{\mathbf{F}} - \dot{e} + T\dot{\eta} - \frac{1}{T}\mathbf{Q} \cdot \mathrm{Grad}T \geq 0. \tag{1.95}$$

The last term in this equation is related to heat flow. Heat flows from the warmer to the colder region of a body. Hence, entropy production by conduction of heat must be nonnegative, i.e.

$$-\frac{1}{T}\mathbf{Q} \cdot \mathrm{Grad}T \geq 0. \tag{1.96}$$

Note. The Fourier's law is widely adopted for heat conduction. In the current configuration, the Fourier law for an isotropic body can be written as

$$\mathbf{q} = -k\mathrm{grad}T, \tag{1.97}$$

where $k > 0$ is the coefficient of thermal conductivity. The heat fluxes should be equal regardless of the reference or current configurations, which gives

$$\oint_{\partial\Omega_0} \mathbf{Q} \cdot \mathbf{N}\mathrm{d}S = \oint_{\partial\Omega} \mathbf{q} \cdot \mathbf{n}\mathrm{d}s. \tag{1.98}$$

Using Nanson's formula (Eq. 1.42), we obtain

$$\mathbf{Q} = J\mathbf{F}^{-1}\mathbf{q}. \tag{1.99}$$

We also have

$$\mathrm{grad}T = \mathrm{Grad}T\,\mathbf{F}^{-1}. \tag{1.100}$$

Combining the previous equations, we can obtain

$$\mathbf{Q} = -kJ\mathbf{F}^{-1}\mathrm{Grad}T\,\mathbf{F}^{-1}. \tag{1.101}$$

Substituting Eq. (1.101) into Eq. (1.96) leads to

$$\frac{kJ}{T}\mathbf{F}^{-1}\mathrm{Grad}T \cdot \mathbf{F}^{-1}\mathrm{Grad}T \geq 0. \tag{1.102}$$

Apparently, this holds true for all conditions with a nonnegative coefficient of thermal conductivity.

We further require that the remaining terms in Eq. (1.95) remain nonnegative, which gives

$$\mathbf{P} : \dot{\mathbf{F}} - \dot{e} + T\dot{\eta} \geq 0. \tag{1.103}$$

This equation is known as the Clausius–Planck inequality.

In continuum mechanics, the Helmholtz free energy density Ψ is also commonly used, which is a Legendre transformation of the internal energy as

$$\Psi = e - T\eta. \tag{1.104}$$

The Clausius–Planck inequality can be rewritten using the Helmholtz free energy density

$$\mathbf{P} : \dot{\mathbf{F}} - \dot{\Psi} - \eta \dot{T} \geq 0. \tag{1.105}$$

Neglecting the thermal effect, we can further obtain

$$\mathbf{P} : \dot{\mathbf{F}} - \dot{\Psi} \geq 0. \tag{1.106}$$

If the process is reversible, no dissipation occurs for the process. It can then be obtained that

$$\mathbf{P} : \dot{\mathbf{F}} - \dot{\Psi} = 0. \tag{1.107}$$

This will be the subject of hyperelasticity, which will be discussed further in Chapter 2.

References

1 Gerhard A. Holzapfel. *Nonlinear Solid Mechanics: A Continuum Approach for Engineering Science*. John Wiley & Sons, Inc., 2000.

2 W Michael Lai, David Rubin, and Erhard Krempl. *Introduction to Continuum Mechanics*. Butterworth-Heinemann, 2009.

3 Morton E Gurtin, Eliot Fried, and Lallit Anand. *The Mechanics and Thermodynamics of Continua*. Cambridge University Press, 2010.

4 Junuthula Narasimha Reddy. *An Introduction to Continuum Mechanics*. Cambridge University Press, 2013.

2

Hyperelastic, Viscoelastic, and Damage Models

Soft materials, such as elastomers and gels, exhibit complex nonlinear mechanical behaviors. Developing the corresponding constitutive models has been a long-term research interest. In this chapter, we introduce several representative models for hyperelastic, viscoelastic, and damage behaviors. For hyperelastic models, we introduce both phenomenological models and micromechanical models. The extended eight-chain model and the Biot chain model are used as examples to show that the recent development of incorporating the entanglement effect into the hyperelastic theories with a considerable improvement on modeling the stress response in multi-axial loading conditions. For the viscoelastic models, the small strain viscoelastic models are briefly reviewed. The chapter introduces four finite deformation viscoelastic models based on internal variables related with either strain or stress. The performance of these models is evaluated through the relaxation modulus upon different stretch levels and the loading–unloading cycles. Finally, the chapter presents several classic damage models and two newly developed damage models for tough elastomers and gels.

2.1 Introduction

Understanding the mechanical behaviors is crucial for developing multi-field coupling models for soft materials. As typical soft materials, elastomers and gels exhibit complex nonlinear responses, including hyperelasticity, viscoelasticity, damage, etc. In soft materials, these nonlinear mechanical behaviors are typically coupled [1], which means that the same materials may exhibit hyperelasticity, viscoelasticity and damage response depending on loading conditions. It is necessary to understand the main characteristics of each type of nonlinear behavior and the general modeling strategies.

The most important feature of hyperelasticity of soft materials is that loading and unloading curves coincide, indicating no energy dissipation. Developing hyperelastic models for rubber-like materials has been an ongoing pursuit, driven by the extensive applications of these materials in various areas. Existing hyperelastic models are developed through either phenomenological approaches or micromechanical frameworks [2–11]. For the phenomenological approach, the strain energy function is proposed to accurately capture experimentally observed stress–strain behaviors. These formulations may rely on strain invariants, exemplified by models such as the neo-Hookean, Mooney–Rivlin, and Gent

Multi-field Modeling of Soft Active Materials: Properties and Design, First Edition. Rui Xiao.

models, or alternatively, they may depend on principal stretches, as seen in the Ogden model [4]. In contrast, micromechanical models are constructed based on understanding the molecular mechanisms governing chain deformation. Prominent examples include the three-chain model [12], the eight-chain model [13], and the microsphere affine model [14]. Recent works [9, 11, 15–17] have incorporated the entanglement effect into the hyperelastic models, which considerably improves the predictions of stress response, especially in biaxial loading conditions.

One of the most prominent features of the mechanical behavior of viscoelastic materials is their time and rate dependence. Specifically, a larger loading rate results in a larger stress response. The viscoelastic models can also generally be categorized into phenomenological models and micromechanically physical models. Phenomenological models are developed based on the rheological models using a spring to represent the elastic response and a dashpot to represent the viscous response [18–21]. These approaches have been extended to finite deformation [1, 22, 23]. Micromechanically based viscoelastic models have also been developed. Vandoolaeghe and Terentjev [24, 25] extended the classic Rouse model and incorporated the entanglement effects. Xiang et al. [16] assumed that the viscoelastic behavior is caused by free chains, considering the non-affine deformation of the molecular chains. Lin et al. [26] proposed a constitutive model for dynamic cross-linked networks based on molecular chain diffusion, integrating the dynamics, diffusion, and deformation of polymer chains.

If damage occurs, the stress response in the preloaded materials is lower than that of the virgin materials, which is known as the Mullins effect [27]. The hyperelastic models can be extended to describe the damage effect based on different approaches [28, 29]. For example, Chagnon et al. [30] developed a continuum damage model for filled rubbers, which can capture the main features of the stress softening behavior. Marckmann et al. [31] proposed a network alteration theory for the Mullins effect, which assumes that the network junctions are broken with deformation, leading to a decrease in chain density and an increase in the number of chain segments. Qi and Boyce [32] adopted a two-phase model for the stress softening, which assumes that the hard phase evolves to the soft phase with deformation. Recent development of the damage model mainly focuses on describing the stress softening effect in tough elastomers and gels, which are mainly extensions of the classic damage models [29, 33, 34].

In this section, we first introduce the constitutive equations for hyperelasticity. Several representative hyperelastic models are also presented. The performances of these models are then evaluated using Treloar's classic data on vulcanized rubbers [35]. This chapter introduces the small strain viscoelastic model serving as the basis for finite deformation viscoelastic models. Four representative viscoelastic models are then presented. The comparison of these models is carried out through the relaxation tests and loading–unloading cycles at a constant loading rate. The procedures of numerical implementation are also briefly discussed. Finally, the chapter presents both classic damage models and several recently developed damage models, which can well predict the damage behavior of tough elastomers and gels.

2.2 Hyperelastic Models

2.2.1 Constitutive Equations of Hyperelastic Materials

A material is referred to as elastic when its mechanical response is solely contingent upon the instantaneous state of deformation. In particular, linear elastic materials represent idealized models that preserve a linear stress–strain relationship throughout deformation. The theory of linear elasticity provides foundation in engineering mechanics, offering the most simple framework for describing the deformations of engineering materials in the small strain regime.

Soft polymeric materials, exemplified by elastomers and gels, often undergo large deformations that surpass the limits of linear elasticity, manifesting pronounced nonlinear mechanical responses. It is necessary to develop hyperelastic models to describe the large deformation behaviors of soft materials. Since the stress state is purely dependent on the deformation state, a strain energy function needs to be defined. Specifically, based on Chapter 1, the Helmholtz free energy function, denoted as Ψ, is chosen throughout the chapter if not specified otherwise.

For kinematics, the deformation gradient is defined as $\mathbf{F}$ and its determinant is denoted as $J = \det(\mathbf{F})$. The right and left Cauchy–Green deformation tensors are defined as $\mathbf{C} = \mathbf{F}^{\mathrm{T}}\mathbf{F}$ and $\mathbf{b} = \mathbf{F}\mathbf{F}^{\mathrm{T}}$.

The first Piola–Kirchhoff stress tensor $\mathbf{P}$ is related to the free energy as

$$\mathbf{P} = \frac{\partial \Psi(\mathbf{F})}{\partial \mathbf{F}}. \tag{2.1}$$

To satisfy material objectivity, the strain energy function must exhibit rotational invariance:

$$\Psi(\mathbf{F}) = \Psi(\mathbf{QF}), \tag{2.2}$$

where $\mathbf{Q}$ is an arbitrary orthogonal tensor.

Applying the polar decomposition theorem, $\mathbf{F} = \mathbf{RU}$, where $\mathbf{R}$ is the rotation tensor and $\mathbf{U}$ is the right stretch tensor, and selecting $\mathbf{Q} = \mathbf{R}^{\mathrm{T}}$, we obtain

$$\Psi(\mathbf{F}) = \Psi(\mathbf{U}). \tag{2.3}$$

Utilizing the right Cauchy–Green deformation tensor, $\mathbf{C} = \mathbf{F}^{\mathrm{T}}\mathbf{F} = \mathbf{U}^2$, the strain energy function can also be expressed as

$$\Psi(\mathbf{F}) = \Psi(\mathbf{C}). \tag{2.4}$$

Consider the time derivative of the strain energy function using the chain rule:

$$\begin{aligned} \dot{\Psi} &= \frac{\partial \Psi(\mathbf{F})}{\partial \mathbf{F}} : \dot{\mathbf{F}} = \operatorname{tr}\left[\left(\frac{\partial \Psi(\mathbf{F})}{\partial \mathbf{F}}\right)^{\mathrm{T}} \cdot \dot{\mathbf{F}}\right] = \operatorname{tr}\left[\left(\frac{\partial \Psi(\mathbf{C})}{\partial \mathbf{C}}\right)^{\mathrm{T}} \cdot \dot{\mathbf{C}}\right] \\ &= \operatorname{tr}\left[\left(\frac{\partial \Psi(\mathbf{C})}{\partial \mathbf{C}}\right)^{\mathrm{T}} \cdot (\dot{\mathbf{F}}^{\mathrm{T}} \cdot \mathbf{F} + \mathbf{F}^{\mathrm{T}} \cdot \dot{\mathbf{F}})\right] = 2\operatorname{tr}\left(\frac{\partial \Psi(\mathbf{C})}{\partial \mathbf{C}} \mathbf{F}^{\mathrm{T}} \cdot \dot{\mathbf{F}}\right). \end{aligned} \tag{2.5}$$

We can then obtain the following relation:

$$\left(\frac{\partial \Psi(\mathbf{F})}{\partial \mathbf{F}}\right)^{\mathrm{T}} = 2\frac{\partial \Psi(\mathbf{C})}{\partial \mathbf{C}} \cdot \mathbf{F}^{\mathrm{T}}. \tag{2.6}$$

Substituting the preceding equation into Eq. (2.1) yields

$$\mathbf{P} = 2\mathbf{F}\frac{\partial \Psi(\mathbf{C})}{\partial \mathbf{C}}. \tag{2.7}$$

Material isotropy implies the same responses when loaded in different directions. For isotropic hyperelasticity, the strain energy function Ψ needs to satisfy

$$\Psi(\mathbf{F}) = \Psi(\mathbf{F}\mathbf{Q}^{\mathrm{T}}). \tag{2.8}$$

Apparently, if the free energy is a function of the principal invariants I_i of $\mathbf{b}$ or $\mathbf{C}$, namely, $\Psi = \Psi(I_1, I_2, I_3)$, this will be a sufficient condition to satisfy Eqs. (2.2) and (2.8).

Applying the chain rule to Eq. (2.7):

$$\frac{\partial \Psi(\mathbf{C})}{\partial \mathbf{C}} = \frac{\partial \Psi}{\partial I_1}\frac{\partial I_1}{\partial \mathbf{C}} + \frac{\partial \Psi}{\partial I_2}\frac{\partial I_2}{\partial \mathbf{C}} + \frac{\partial \Psi}{\partial I_3}\frac{\partial I_3}{\partial \mathbf{C}}. \tag{2.9}$$

The derivatives of the invariants with respect to $\mathbf{C}$ are

$$\begin{aligned}
\frac{\partial I_1}{\partial \mathbf{C}} &= \frac{\partial \mathrm{tr}\mathbf{C}}{\partial \mathbf{C}} = \frac{\partial(\mathbf{I} : \mathbf{C})}{\partial \mathbf{C}} = \mathbf{I},\\
\frac{\partial I_2}{\partial \mathbf{C}} &= \frac{1}{2}\left(2\mathrm{tr}(\mathbf{C})\mathbf{I} - \frac{\partial \mathrm{tr}(\mathbf{C}^2)}{\partial \mathbf{C}}\right) = I_1\mathbf{I} - \mathbf{C},\\
\frac{\partial I_3}{\partial \mathbf{C}} &= \frac{\partial \det(\mathbf{C})}{\partial \mathbf{C}} = \det\mathbf{C}\cdot\mathbf{C}^{-\mathrm{T}} = I_3\mathbf{C}^{-1}.
\end{aligned} \tag{2.10}$$

Substituting Eqs. (2.9) and (2.10) into Eq. (2.7) yields

$$\mathbf{P} = 2\mathbf{F}\left(\left(\frac{\partial \Psi}{\partial I_1} + I_1\frac{\partial \Psi}{\partial I_2}\right)\mathbf{I} - \frac{\partial \Psi}{\partial I_2}\mathbf{C} + I_3\frac{\partial \Psi}{\partial I_3}\mathbf{C}^{-1}\right). \tag{2.11}$$

The corresponding second Piola–Kirchhoff stress ($\mathbf{S}$) and Cauchy stress ($\boldsymbol{\sigma}$) can be represented as

$$\mathbf{S} = 2\left(\left(\frac{\partial \Psi}{\partial I_1} + I_1\frac{\partial \Psi}{\partial I_2}\right)\mathbf{I} - \frac{\partial \Psi}{\partial I_2}\mathbf{C} + I_3\frac{\partial \Psi}{\partial I_3}\mathbf{C}^{-1}\right), \tag{2.12a}$$

$$\boldsymbol{\sigma} = \frac{2}{J}\left(\left(\frac{\partial \Psi}{\partial I_1} + I_1\frac{\partial \Psi}{\partial I_2}\right)\mathbf{b} - \frac{\partial \Psi}{\partial I_2}\mathbf{b}^2 + I_3\frac{\partial \Psi}{\partial I_3}\mathbf{I}\right). \tag{2.12b}$$

2.2.1.1 Incompressible Hyperelastic Materials

For incompressible materials ($J = \det\mathbf{F} = 1$), we introduce a Lagrange multiplier p to reinforce incompressibility. Thus, the free energy can be represented as

$$\Psi = \Psi(I_1, I_2) - p(J - 1). \tag{2.13}$$

In this condition, the stress can be represented as

$$\mathbf{P} = 2\mathbf{F}\left(\left(\frac{\partial \Psi}{\partial I_1} + I_1\frac{\partial \Psi}{\partial I_2}\right)\mathbf{I} - \frac{\partial \Psi}{\partial I_2}\mathbf{C}\right) - p\mathbf{F}^{-\mathrm{T}}, \tag{2.14a}$$

$$\mathbf{S} = 2\left(\left(\frac{\partial \Psi}{\partial I_1} + I_1\frac{\partial \Psi}{\partial I_2}\right)\mathbf{I} - \frac{\partial \Psi}{\partial I_2}\mathbf{C}\right) - p\mathbf{C}^{-1}, \tag{2.14b}$$

$$\boldsymbol{\sigma} = 2\left(\frac{\partial \Psi}{\partial I_1} + I_1\frac{\partial \Psi}{\partial I_2}\right)\mathbf{b} - 2\frac{\partial \Psi}{\partial I_2}\mathbf{b}^2 - p\mathbf{I}. \tag{2.14c}$$

If the free energy is a function of the principal stretch λ_i, $i = 1, 2, 3$, the energy function for the incompressible condition can be represented as

$$\Psi = \Psi(\lambda_1, \lambda_2, \lambda_3) - p(J - 1). \tag{2.15}$$

The principal stress can then be represented as

$$P_i = \frac{\partial \Psi}{\partial \lambda_i} - \frac{1}{\lambda_i}p, \tag{2.16a}$$

$$S_i = \frac{1}{\lambda_i}\frac{\partial \Psi}{\partial \lambda_i} - \frac{1}{\lambda_i^2}p, \tag{2.16b}$$

$$\sigma_i = \lambda_i\frac{\partial \Psi}{\partial \lambda_i} - p. \tag{2.16c}$$

2.2.1.2 Compressible Hyperelastic Materials

All materials exhibit a certain degree of compressibility. Moreover, in finite element implementations, incompressible hyperelastic models often encounter convergence issues such as volumetric locking. Thus, it is necessary to extend the above formulation to the quasi-incompressible condition.

The deformation gradient $\mathbf{F}$ is decomposed into isochoric and volumetric components:

$$\mathbf{F} = J^{1/3}\bar{\mathbf{F}}. \tag{2.17}$$

The isochoric Cauchy–Green deformation tensor can be represented as

$$\bar{\mathbf{C}} = \bar{\mathbf{F}}^{\mathrm{T}}\bar{\mathbf{F}}, \quad \bar{\mathbf{b}} = \bar{\mathbf{F}}\bar{\mathbf{F}}^{\mathrm{T}}. \tag{2.18}$$

The invariant of the isochoric Cauchy–Green deformation tensor can be represented as $\bar{I}_i$, $i = 1, 2, 3$.

The free energy has contributions from both the isochoric and volumetric parts:

$$\Psi(\mathbf{C}) = \Psi_{\mathrm{iso}}(\bar{\mathbf{C}}) + \Psi_{\mathrm{vol}}(J). \tag{2.19}$$

We first give the derivative of $\bar{\mathbf{C}}$ with respect to $\mathbf{C}$, which gives

$$\begin{aligned}\frac{\partial \bar{\mathbf{C}}}{\partial \mathbf{C}} &= \frac{\partial (J^{-2/3}\mathbf{C})}{\partial \mathbf{C}} \\ &= J^{-2/3}\left(\mathbb{I} + J^{2/3}\mathbf{C} \otimes \frac{\partial J^{-2/3}}{\partial \mathbf{C}}\right) \\ &= J^{-2/3}\left(\mathbb{I} - \frac{1}{3}\mathbf{C} \otimes \mathbf{C}^{-1}\right),\end{aligned} \tag{2.20}$$

where $\mathbb{I}$ denotes the fourth-order identity tensor.

The isochoric second Piola–Kirchhoff stress can be represented as

$$\begin{aligned}\mathbf{S}_{\text{iso}} &= 2\frac{\partial \Psi_{\text{iso}}(\bar{\mathbf{C}})}{\partial \mathbf{C}} \\ &= 2\left(\frac{\partial \Psi_{\text{iso}}}{\partial \bar{\mathbf{C}}} : \frac{\partial \bar{\mathbf{C}}}{\partial \mathbf{C}}\right) \\ &= 2J^{-2/3}\left[\frac{\partial \Psi_{\text{iso}}}{\partial \bar{\mathbf{C}}} : \left(\mathbb{I} - \frac{1}{3}\mathbf{C} \otimes \mathbf{C}^{-1}\right)\right].\end{aligned} \tag{2.21}$$

The Cauchy stress can then be represented as

$$\boldsymbol{\sigma}_{\text{iso}} = \frac{1}{J}\mathbf{FSF}^{\text{T}} = \frac{2}{J}\bar{\mathbf{F}}\left[\frac{\partial \Psi_{\text{iso}}}{\partial \bar{\mathbf{C}}} : \left(\mathbb{I} - \frac{1}{3}\mathbf{C} \otimes \mathbf{C}^{-1}\right)\right]\bar{\mathbf{F}}^{\text{T}}. \tag{2.22}$$

The volumetric second Piola–Kirchhoff stress and the Cauchy stress can be represented as

$$\mathbf{S}_{\text{vol}} = 2\frac{\partial \Psi_{\text{vol}}(J)}{\partial \mathbf{C}} = J\frac{\partial \Psi_{\text{vol}}(J)}{\partial J}\mathbf{C}^{-1}, \tag{2.23a}$$

$$\boldsymbol{\sigma}_{\text{vol}} = \frac{1}{J}\mathbf{FSF}^{\text{T}} = \frac{\partial \Psi_{\text{vol}}(J)}{\partial J}\mathbf{I}. \tag{2.23b}$$

Two common expressions are widely adopted to represent the volumetric free energy density $\Psi_{\text{vol}}(J)$:

$$\Psi^1_{\text{vol}}(J) = \frac{\kappa}{4}(J^2 - 2\ln J - 1), \tag{2.24a}$$

$$\Psi^2_{\text{vol}}(J) = \frac{\kappa}{2}(J - 1)^2, \tag{2.24b}$$

where κ is the bulk modulus.

Combining the preceding two equations leads to the following form of the volumetric Cauchy stress:

$$\boldsymbol{\sigma}^1_{\text{vol}} = \frac{\kappa}{2}\left(J - \frac{1}{J}\right)\mathbf{I}, \tag{2.25a}$$

$$\boldsymbol{\sigma}^2_{\text{vol}} = \kappa(J - 1)\mathbf{I}. \tag{2.25b}$$

2.2.2 Hyperelastic Models

Hyperelastic models can be classified into two main categories: phenomenological models and micromechanical models. Phenomenological hyperelastic models aim to describe material behavior by constructing polynomial functions of strain invariants (such as the Mooney–Rivlin model) or principal stretch (such as the Ogden model). The merit of these models lies in their mathematical simplicity, coupled with the ability to accurately predict experimental data. Micromechanical models are developed based on the relation between the microscopic deformation of polymer chains and macroscopic deformation, which are mainly derived using statistical mechanics. These models attribute the hyperelastic behavior of soft materials to change in the conformational entropy of molecular chains.

Based on different micro–macro transition, classic micromechanical models include three-chain, four-chain, eight-chain, and full-network models. Recent developments in micromechanical models include the entanglement effect mainly based on the tube model. In this part, we will introduce several classic hyperelastic models as well as the newly developed model incorporating the entanglement effect.

2.2.2.1 Neo-Hookean Model

Among all hyperelastic constitutive models, the neo-Hookean model stands as the most fundamental model. In fact, many hyperelastic models can be reduced to the neo-Hookean model under certain conditions. Furthermore, the neo-Hookean model can also be derived from a statistical perspective rooted in the molecular chain structure. This model was initially proposed by Treloar [36], with the free energy function as

$$\Psi = \frac{1}{2} G (I_1 - 3), \tag{2.26}$$

where G is the shear modulus.

In the incompressible condition, the Cauchy stress can be obtained based on Eqs. (2.14c) and (2.26) as

$$\boldsymbol{\sigma} = G\mathbf{b} - p\mathbf{I}. \tag{2.27}$$

For the quasi-incompressible condition, the free energy can be represented as

$$\Psi = \frac{1}{2} G (\bar{I}_1 - 3) + \frac{\kappa}{2} (J - 1)^2. \tag{2.28}$$

The Cauchy stress can then be calculated as

$$\boldsymbol{\sigma} = G\left(\bar{\mathbf{b}} - \frac{1}{3}\bar{I}_1\mathbf{I}\right) + \kappa(J - 1)\mathbf{I}. \tag{2.29}$$

2.2.2.2 Mooney–Rivlin Model

Mooney [2] constructed a free energy function, which is linearly dependent on the first and second strain tensor invariants (I_1, I_2), represented as

$$\Psi = C_{10}(I_1 - 3) + C_{01}(I_2 - 3), \tag{2.30}$$

where C_{10} and C_{01} are material constants.

In the incompressible condition, the Cauchy stress can be further obtained based on Eqs. (2.14c) and (2.30) as

$$\boldsymbol{\sigma} = 2C_{10}\mathbf{b} + 2C_{01}(I_1\mathbf{I} - \mathbf{b})\mathbf{b} - p\mathbf{I}. \tag{2.31}$$

The corresponding quasi-incompressible condition can be extended following the similar approach as the neo-Hookean model, which will not be further elaborated here.

In an effort to address the limitations of the Mooney model, Rivlin [37] extended the strain energy function into a series polynomial form of I_1 and I_2, thereby developing the generalized Mooney–Rivlin model:

$$\Psi = \sum_{i+j\geq 1} C_{ij}(I_1 - 3)^i (I_2 - 3)^j, \tag{2.32}$$

where C_{ij} are higher-order material constants. By truncating the polynomial order, this model can be reduced to various classic forms [38].

2.2.2.3 Ogden Model

The Ogden model, based on principal stretch, has the following form [4]:

$$\Psi(\lambda_1, \lambda_2, \lambda_3) = \sum_{i=1}^{n} \frac{\mu_i}{\alpha_i}(\lambda_1^{\alpha_i} + \lambda_2^{\alpha_i} + \lambda_3^{\alpha_i} - 3), \tag{2.33}$$

where α_i is a dimensionless coefficient to be determined and μ_i is a material constant related to the shear modulus. The shear modulus in the small region is related with α_i and μ_i

$$2G = \sum_{i=1}^{n} \mu_i \alpha_i. \tag{2.34}$$

The principal Cauchy stress can be calculated based on Eqs. (2.33) and (2.16c):

$$\sigma_j = \sum_{i=1}^{n} \mu_i \lambda_j^{\alpha_i} - p, \quad j = 1, 2, 3. \tag{2.35}$$

2.2.2.4 Gent Model

Gent [3] proposed a hybrid model, which combines phenomenological approaches with the material's microstructure. This model accounts for the stretch limit of the molecular chains, with the free energy density function expressed as

$$\Psi = -\frac{G}{2} J_m \ln\left(1 - \frac{I_1 - 3}{J_m}\right), \tag{2.36}$$

where J_m is a material constant with $0 \leq (I_1 - 3)/J_m < 1$. As the molecular chains approach their limiting stretch ratio, the stress reaches infinity.

The Cauchy stress has the following form for the incompressible condition:

$$\boldsymbol{\sigma} = G \frac{1}{1 - (I_1 - 3)/J_m} \mathbf{b} - p\mathbf{I}. \tag{2.37}$$

This model characterizes the strain stiffening effect in rubbery materials in the large deformation regime, making it a representative model for describing the mechanical behavior of these materials in the whole deformation region.

2.2.2.5 Eight-chain Model

James et al. [39] developed the three-chain model, where the probability density of the chain end-to-end distance is described by the Langevin statistical function, assuming that all chains are aligned uniformly along the principal axes within a unit cell. Flory [40] proposed the four-chain model, where four molecular chains extend from the vertices of a regular tetrahedron and cross-link at the center, with chains deforming non-affinely during macroscopic deformation. Arruda and Boyce [13] developed the eight-chain model, wherein eight chains extend from the vertices of a cubic unit cell and cross-link at the center. The symmetric structure ensures that the cross-link point remains at the center during deformation, with all chains undergoing identical stretch.

We use eight-chain model as an example to show the general procedures of developing micromechanical hyperelastic models. The Langevin statistics is used to represent the free energy density of a single chain [13]:

$$\psi^{c} = kTN\left(\frac{\lambda}{\sqrt{N}}\beta + \ln\frac{\beta}{\sinh\beta}\right) + \text{constant}, \quad \beta = \mathcal{L}^{-1}\left(\frac{\lambda}{\sqrt{N}}\right), \tag{2.38}$$

where k is the Boltzmann constant, T is the temperature, N is the number of chain segments, and λ is the stretch ratio. The stretch ratio λ is related to the end-to-end distance of the chain r as $\lambda = r/\left(\sqrt{N}b\right)$, where b is the Kuhn length. The function $\mathcal{L}^{-1}(\cdot)$ is the inverse Langevin function with the Langevin function defined as $\mathcal{L}(x) = \coth(x) - 1/x$.

For the eight-chain model, the free energy is expressed as

$$\Psi = nkTN\left(\frac{\bar{\lambda}}{\sqrt{N}}\beta + \ln\frac{\beta}{\sinh\beta}\right) + \Psi_0, \quad \bar{\lambda} = \sqrt{\frac{\bar{I}_1}{3}}, \quad \beta = \mathcal{L}^{-1}\left(\frac{\bar{\lambda}}{\sqrt{N}}\right), \tag{2.39}$$

where n represents the number of polymer chains per unit volume. The shear modulus can be written as $G = nkT$.

The Cauchy stress can be calculated as

$$\boldsymbol{\sigma} = \frac{G}{3}\frac{\sqrt{N}}{\bar{\lambda}}\beta\mathbf{b} - p\mathbf{I}. \tag{2.40}$$

As discussed in Xiao et al. [41], the following function proposed by Kroeger [42] gives a good approximation for the inverse Langevin function:

$$\mathcal{L}^{-1}(x) = \frac{3x}{(1-x^2)(1+\frac{1}{2}x^2)}. \tag{2.41}$$

2.2.2.6 Extended Eight-chain Model

Existing models often fail to accurately predict stress responses under biaxial loading conditions. This fundamental limitation may be caused by inadequate consideration of the entanglement effect, which suggests that the deformation of chains is also affected by their surrounding chains. This can be addressed through explicit modeling of topological constraints. Thus, the stress response can be attributed to two components: the cross-linked network and the entanglement network, as illustrated in Figure 2.1.

The tube model has been successfully applied to modeling the entanglement effect. Doi et al. [43] introduced a virtual tube paradigm where polymer chains are confined within

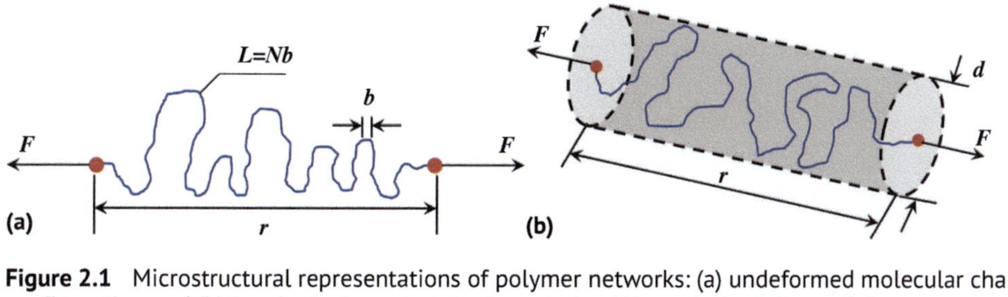

Figure 2.1 Microstructural representations of polymer networks: (a) undeformed molecular chain configuration and (b) topological constraints characterized through the tube model.

conceptual tubes that restrict their kinematic motion. The entanglement intensity evolves with tube cross-sectional contraction, quantified through a linear relationship between entanglement free energy density and tube area contraction ratio:

$$\psi^{\mathrm{e}} = \alpha kTN\left(\frac{b}{d_0}\right)^2 \upsilon, \tag{2.42}$$

where α denotes the entanglement intensity constant, d_0 is the initial tube diameter, b is the Kuhn length, and υ is the ratio of deformed-to-initial tube cross-sectional areas.

The remaining task is determining the micro–macro transition. Khiêm and Itskov [9] and Shen et al. [11] have adopted the network averaging approach to build the relation between the deformation of chains in the micro-scale to the macro-scale deformation, which gives

$$\bar{\lambda} = \sqrt{\frac{I_1}{3}}, \quad \bar{\upsilon} = \sqrt{\frac{I_2}{3}}. \tag{2.43}$$

Thus, the free energy of the system can be represented as

$$\Psi = \Psi_{\mathrm{c}} + \Psi_{\mathrm{e}}, \tag{2.44a}$$

$$\Psi_{\mathrm{c}} = G_{\mathrm{c}} N\left(\frac{\bar{\lambda}}{\sqrt{N}}\beta + \ln\frac{\beta}{\sinh\beta}\right), \tag{2.44b}$$

$$\Psi_{\mathrm{e}} = 3G_{\mathrm{e}}\sqrt{\frac{I_2}{3}}, \tag{2.44c}$$

where G_{c} is the shear modulus of the cross-linked network and $G_{\mathrm{e}} = \alpha nkTN(b/d_0)^2/3$ is the shear modulus of the entanglement part.

As shown in Eqs. (2.39) and (2.44), the free energy function for the cross-linked part of the above model is the same as the eight-chain model. Thus, the above model is named as the extended eight-chain model [17].

The Cauchy stress can be further calculated as

$$\boldsymbol{\sigma} = \frac{G_{\mathrm{c}}}{3}\frac{\sqrt{N}}{\sqrt{I_1/3}}\beta\mathbf{b} + \frac{G_{\mathrm{e}}}{\sqrt{I_2/3}}(I_1\mathbf{b} - \mathbf{b}^2) - p\mathbf{I}. \tag{2.45}$$

2.2.2.7 Biot Chain Model

A new non-affine micro–macro transition was proposed by Amores et al. [44] and Zhan et al. [45], which can be represented as

$$\lambda_n = \mathbf{nUn}, \tag{2.46}$$

where $\mathbf{n}$ is the direction and $\mathbf{U}$ is the right stretch tensor. As discussed in Zhan et al. [45], this micro–macro transition can be derived based on the fact that the macro Biot stress equals to the average of the micro Biot stress.

The above approach is further combined with the microsphere model. The free energy density can be obtained through integrating the free energy density of each direction ($\psi(\lambda_n)$) over a sphere surface:

$$\Psi = \frac{1}{S}\int n_0\psi(\lambda_n)\mathrm{d}s, \tag{2.47}$$

with n_0 being the number of chain density per unit volume and S being the surface area of the unit sphere.

For a Gaussian chain distribution, the free energy of a single chain can be represented as

$$\psi(\lambda) = \frac{3}{2}kT\lambda^2 + \text{constant}, \tag{2.48}$$

where k is the Boltzmann constant, T is the absolute temperature, and λ is the stretch ratio.

Combining Eqs. (2.46)–(2.48) yields

$$\begin{aligned}\Psi &= \frac{1}{7}G\left((\text{tr}\mathbf{U})^2 + 2\text{tr}\mathbf{U}^2 - 15\right)\\ &= \frac{1}{7}G\left(3\lambda_1^2 + 3\lambda_2^2 + 3\lambda_3^2 + 2\lambda_1\lambda_2 + 2\lambda_2\lambda_3 + 2\lambda_1\lambda_3 - 15\right),\end{aligned} \tag{2.49}$$

where $G = 7n_0kT/10$.

If strain hardening needs to be considered, the non-Gaussian chain model needs to be used. Specifically, the free energy of a single chain based on Langevin chain statistics can be adopted with the form as

$$\psi = kTN\left(\frac{\lambda\beta}{\sqrt{N}} + \ln\frac{\beta}{\sinh\beta}\right) + \text{constant}. \tag{2.50}$$

This is the same as Eq. (2.38).

Combining Eqs. (2.46), (2.47), and (2.50) yields the following form of the free energy density:

$$\Psi = \frac{10}{7}GN\sum_i\left(\frac{\tilde{\lambda}_i\beta_i}{\sqrt{N}} + \ln\frac{\beta_i}{\sinh\beta_i}\right)\omega_i + \text{constant}, \quad \tilde{\lambda}_i = \mathbf{n}_i\mathbf{U}\mathbf{n}_i, \tag{2.51}$$

where $\mathbf{n}_i$ is the direction of discrete chains, ω_i is the direction weight ratio, $\beta_i = \mathcal{L}^{-1}(\tilde{\lambda}_i/\sqrt{N})$, and $\tilde{\lambda}_i$ is the stretch in each discrete direction.

2.2.3 Results

In this part, we evaluate the performance of the above models using the Treloar's classic experimental data [35], which includes the stress response of vulcanized rubber loaded in uniaxial tension, pure shear, and equibiaxial stretching conditions. The fitting methodology employed here simultaneously incorporates all experimental deformation modes. As shown in Figure 2.2, the neo-Hookean model can only capture the stress response in the small to median deformation regime of the uniaxial test. The Mooney–Rivlin can effectively capture the stress in biaxial tests, while it fails to reproduce the strain stiffening effect. In contrast, the Gent model and the eight-chain model can well capture the stress response in all the deformation region for the uniaxial loading condition. However, both models exhibit significant deviations when predicting equibiaxial responses. This predictive inadequacy becomes more pronounced when the model parameters are calibrated using single deformation modes (e.g. uniaxial or equibiaxial data alone).

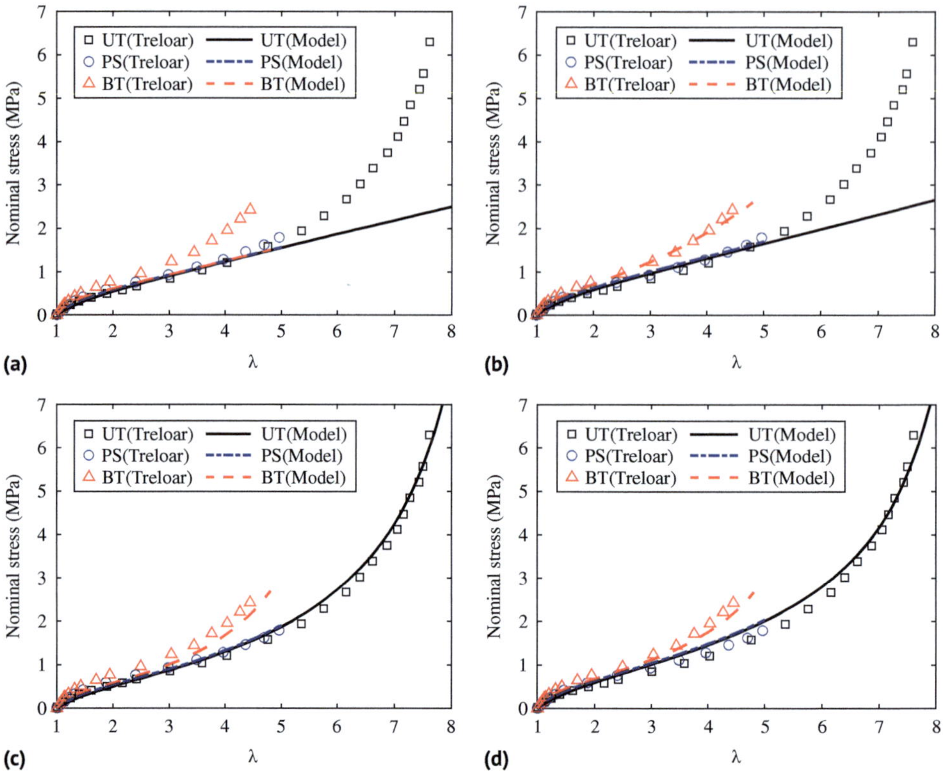

Figure 2.2 Predictions of different models against Treloar's experimental data: (a) the neo-Hookean model with $G = 0.312$ MPa, (b) the Mooney–Rivlin model with $C_{10} = 0.1655$ MPa, $C_{01} = 0.0046$ MPa, (c) the Gent model with $G = 0.278$ MPa and $J_m = 85.23$, and (d) the eight-chain model with $G = 0.328$ MPa and $N = 28.6$.

We further apply the extended eight-chain model to predict the Treloar's data. Again, the model parameters are fitted to all three loading modes to obtain the minimum average relative error [46]. As shown in Figure 2.3, the extended eight-chain model can capture the stress response in all three loading modes, which shows a considerable improvement compared with the classic hyperelastic models. As discussed in Shen et al. [11], the entanglement effect is more pronounced in the equibiaxial loading conditions than that in the uniaxial loading conditions.

The above results show that the hyperelastic model with the entanglement effect can well describe the complex stress response in both uniaxial and biaxial loading conditions. However, as discussed in Zhu et al. [46], the model parameters need to be obtained through fitting to uniaxial and biaxial data simultaneously. The Biot chain model developed in Zhan et al. [45] can overcome this limitation. As shown in Figure 2.4, the two model parameters can either be obtained from the uniaxial data or equibiaxial data. The model can then well predict the stress response in the other two loading modes.

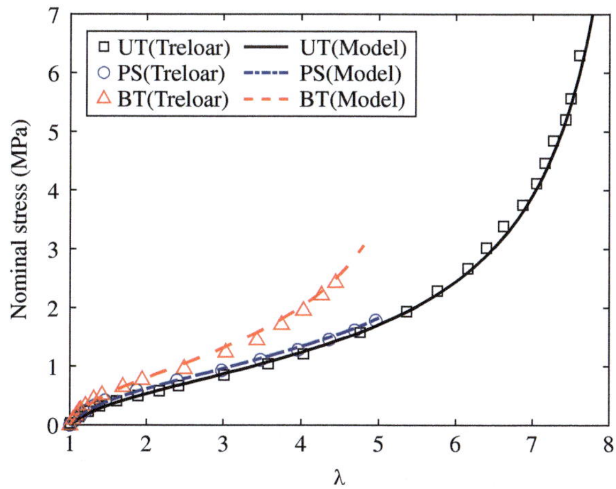

Figure 2.3 Predictive performance of the extended eight-chain model against Treloar's data with parameters $G_c = 0.263$ MPa, $N = 25.64$ and $G_e = 0.077$ MPa.

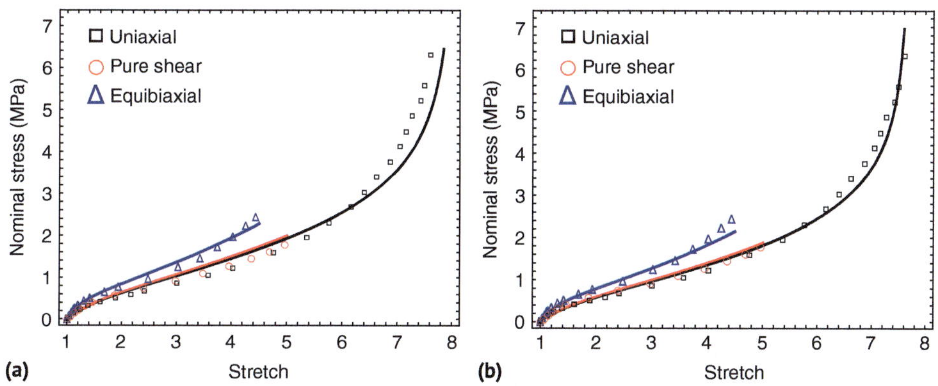

Figure 2.4 Predictions Treloar's experimental data using the Biot chain model with (a) $G = 0.347$ MPa and $N = 61.92$ using the uniaxial data to calibrate the model parameter and (b) $G = 0.376$ MPa and $N = 67.69$ using the uniaxial data to calibrate the model parameter.

2.3 Viscoelastic Models

In this part, we first introduce the one-dimensional small strain viscoelastic model. This model is the basis of the finite deformation viscoelastic models. In fact, all the four finite deformation viscoelastic models introduced in this section can be reduced to the small strain model.

2.3.1 One-dimensional Small Strain Viscoelastic Models

For a viscoelastic material, we can use an elastic spring to represent the elastic response, while a dashpot is typically adopted to describe the viscous response. Different arrangement of spring and dashpot yields various viscoelastic models. As shown in Figure 2.5, the Maxwell model is composed of a spring in series with a dashpot. The generalized Maxwell model is composed of a spring to describe the equilibrium elastic response and multiple

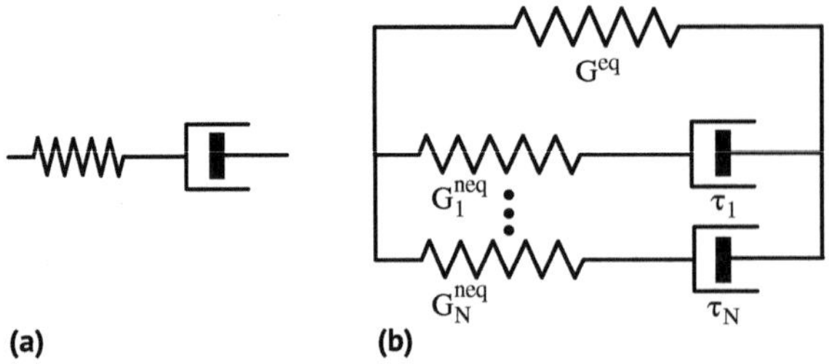

Figure 2.5 Rheological representative of viscoelastic models: (a) the Maxwell model and (b) the generalized Maxwell model.

Maxwell elements arranged in parallel to represent the viscoelastic response, which is adequate to describe the viscoelastic effect of various polymeric materials.

In each Maxwell element, the total strain of the spring $\varepsilon_k^{\mathrm{e}}$ and the dashpot $\varepsilon_k^{\mathrm{v}}$ should be equal to the strain ε in the elastic branch, which yields

$$\varepsilon = \varepsilon_k^{\mathrm{e}} + \varepsilon_k^{\mathrm{v}}, \quad k = 1, \ldots, N, \tag{2.52}$$

where N is the total number of Maxwell elements.

The total stress is given by

$$\sigma = \sigma^{\mathrm{eq}} + \sum_k^N \sigma_k^{\mathrm{neq}}, \tag{2.53a}$$

$$\sigma^{\mathrm{eq}} = G^{\mathrm{eq}}\varepsilon, \tag{2.53b}$$

$$\sigma_k^{\mathrm{neq}} = G_k^{\mathrm{neq}}\varepsilon_k^{\mathrm{e}}, \tag{2.53c}$$

where G^{eq} is the equilibrium shear modulus and G_k^{neq} is the nonequilibrium shear modulus of the kth Maxwell element.

The following linear evolution equation is adopted for $\varepsilon_k^{\mathrm{v}}$:

$$\dot{\varepsilon}_k^{\mathrm{v}} = \frac{\sigma_k^{\mathrm{neq}}}{\eta_k}, \quad \varepsilon_k^{\mathrm{v}}(t=0) = 0, \tag{2.54}$$

where η_k is the viscosity. The relaxation time is defined as $\tau_k = \eta_k / G_k^{\mathrm{neq}}$. Thus, the preceding equation can be written as

$$\dot{\varepsilon}_k^{\mathrm{v}} = \frac{\varepsilon - \varepsilon_k^{\mathrm{v}}}{\tau_k}. \tag{2.55}$$

Combining Eqs. (2.52), (2.53c), and (2.55) and eliminating the internal variables $\varepsilon_k^{\mathrm{e}}$ and $\varepsilon_k^{\mathrm{v}}$ leads to the following evolution equation for the nonequilibrium stress σ_k^{neq} as

$$\dot{\sigma}_k^{\mathrm{neq}} + \frac{\sigma_k^{\mathrm{neq}}}{\tau_k} = G_k^{\mathrm{neq}}\dot{\varepsilon}. \tag{2.56}$$

Thus, we have two ways to present the generalized Maxwell model. For the first approach, the model is based on the strain decomposition, which consists of Eqs. (2.52), (2.53), and (2.55). The second approach employs the evolution of the nonequilibrium stress with Eqs. (2.53a), (2.53b), and (2.56).

2.3.2 Finite Deformation Viscoelastic Models

In this section, we introduce finite deformation viscoelastic models. The complexity lies in different types of stress and strain due to varied configurations. In the following, we introduce two finite deformation models using internal variables related with strain. The finite deformation models based on the second Piola–Kirchhoff stress and the Cauchy stress are also introduced.

2.3.2.1 Multiplicative Decomposition Model

Based on the idea of multiplicative decomposition, Reese and Govindjee [47] developed a finite deformation viscoelastic model. The deformation gradient is multiplicatively decomposed into elastic and viscous parts as

$$\mathbf{F} = \mathbf{F}_k^{\mathrm{e}}\mathbf{F}_k^{\mathrm{v}}, \quad k = 1, \ldots, N. \tag{2.57}$$

The left and right Cauchy–Green elastic deformation tensors are defined as

$$\mathbf{b}_k^{\mathrm{e}} = \mathbf{F}_k^{\mathrm{e}}\mathbf{F}_k^{\mathrm{e}^{\mathrm{T}}}, \quad \mathbf{C}_k^{\mathrm{e}} = \mathbf{F}_k^{\mathrm{e}^{\mathrm{T}}}\mathbf{F}_k^{\mathrm{e}}. \tag{2.58}$$

To represent the inherent difference in the time-dependent behavior of the volumetric and deviatoric response, we also split the deformation into distortional and volumetric parts as

$$\bar{\mathbf{b}}_k^{\mathrm{e}} = J_k^{\mathrm{e}^{-2/3}}\mathbf{b}_k^{\mathrm{e}}, \quad \bar{\mathbf{C}}_k^{\mathrm{e}} = J_k^{\mathrm{e}^{-2/3}}\mathbf{C}_k^{\mathrm{e}}, \tag{2.59}$$

where $J_k^{\mathrm{e}} = \det(\mathbf{F}_k^{\mathrm{e}})$ is the elastic volumetric deformation ratio of the kth relaxation process.

The viscous velocity gradients and its symmetric part are defined as

$$\mathbf{L}_k^{\mathrm{v}} = \dot{\mathbf{F}}_k^{\mathrm{v}}\mathbf{F}_k^{\mathrm{v}^{-1}}, \quad \mathbf{D}_k^{\mathrm{v}} = \frac{\mathbf{L}_k^{\mathrm{v}} + \mathbf{L}_k^{\mathrm{v}^{\mathrm{T}}}}{2}. \tag{2.60}$$

The Helmholtz free energy density is also assumed to be composed of an equilibrium part as a function of the right Cauchy deformation tensor $\mathbf{C}$ and multiple nonequilibrium parts as a function the elastic deformation gradient $\mathbf{C}_k^{\mathrm{e}}$, which is represented as

$$\Psi = \Psi^{\mathrm{eq}}(\mathbf{C}) + \sum_k^N \Psi_k^{\mathrm{neq}}(\mathbf{C}_k^{\mathrm{e}}). \tag{2.61}$$

The rate of Helmholtz free energy density is then expressed as

$$\begin{aligned}
\dot{\Psi} &= 2\frac{\partial \Psi^{\mathrm{eq}}}{\partial \mathbf{C}} : \frac{1}{2}\dot{\mathbf{C}} + \sum_k^N \frac{\partial \Psi_k^{\mathrm{neq}}}{\partial \mathbf{C}_k^{\mathrm{e}}} : \dot{\mathbf{C}}_k^{\mathrm{e}} \\
&= 2\frac{\partial \Psi^{\mathrm{eq}}}{\partial \mathbf{C}} : \frac{1}{2}\dot{\mathbf{C}} + \sum_k^N \frac{\partial \Psi_k^{\mathrm{neq}}}{\partial \mathbf{C}_k^{\mathrm{e}}} : \frac{\partial \mathbf{C}_k^{\mathrm{e}}}{\partial \mathbf{C}} : \dot{\mathbf{C}} + \sum_k^N \frac{\partial \Psi_k^{\mathrm{neq}}}{\partial \mathbf{C}_k^{\mathrm{e}}} : \frac{\partial \mathbf{C}_k^{\mathrm{e}}}{\partial \mathbf{F}_k^{\mathrm{v}}} : \dot{\mathbf{F}}_k^{\mathrm{v}} \\
&= 2\frac{\partial \Psi^{\mathrm{eq}}}{\partial \mathbf{C}} : \frac{1}{2}\dot{\mathbf{C}} + \sum_k^N 2\mathbf{F}_k^{\mathrm{v-1}}\frac{\partial \Psi_k^{\mathrm{neq}}}{\partial \mathbf{C}_k^{\mathrm{e}}}\mathbf{F}_k^{\mathrm{v-T}} : \frac{1}{2}\dot{\mathbf{C}} - \sum_k^N \frac{\partial \Psi_k^{\mathrm{neq}}}{\partial \mathbf{C}_k^{\mathrm{e}}} : (\mathbf{L}_k^{\mathrm{vT}}\mathbf{C}_k^{\mathrm{e}} + \mathbf{C}_k^{\mathrm{e}}\mathbf{L}_k^{\mathrm{v}}) \\
&= 2\frac{\partial \Psi^{\mathrm{eq}}}{\partial \mathbf{C}} : \frac{1}{2}\dot{\mathbf{C}} + \sum_k^N 2\mathbf{F}_k^{\mathrm{v-1}}\frac{\partial \Psi_k^{\mathrm{neq}}}{\partial \mathbf{C}_k^{\mathrm{e}}}\mathbf{F}_k^{\mathrm{v-T}} : \frac{1}{2}\dot{\mathbf{C}} - \sum_k^N 2\mathbf{C}_k^{\mathrm{e}}\frac{\partial \Psi_k^{\mathrm{neq}}}{\partial \mathbf{C}_k^{\mathrm{e}}} : \mathbf{D}_k^{\mathrm{v}}.
\end{aligned} \tag{2.62}$$

As shown in Chapter 1, the Clausius–Duhem inequality is written as

$$\mathbf{S} : \frac{1}{2}\dot{\mathbf{C}} - \dot{\Psi} \geq 0. \tag{2.63}$$

Combining Eqs. (2.62) and (2.63) leads to the following form for the entropy inequality:

$$\left(\mathbf{S} - 2\frac{\partial \Psi^{\mathrm{eq}}}{\partial \mathbf{C}} - 2\sum_{k}^{N} \mathbf{F}_k^{\mathrm{v}\text{-}1}\frac{\partial \Psi_k^{\mathrm{neq}}}{\partial \mathbf{C}_k^{\mathrm{e}}}\mathbf{F}_k^{\mathrm{v}\text{-}\mathrm{T}}\right) : \frac{1}{2}\dot{\mathbf{C}} + \sum_{k}^{N} 2\mathbf{C}_k^{\mathrm{e}}\frac{\partial \Psi_k^{\mathrm{neq}}}{\partial \mathbf{C}_k^{\mathrm{e}}} : \mathbf{D}_k^{\mathrm{v}} \geq 0. \tag{2.64}$$

The preceding equation (2.64) should hold at the equilibrium condition, which leads to the following constitutive relations:

$$\mathbf{S} = 2\frac{\partial \Psi^{\mathrm{eq}}}{\partial \mathbf{C}} + 2\sum_{k}^{N} \mathbf{F}_k^{\mathrm{v}\text{-}1}\frac{\partial \Psi_k^{\mathrm{neq}}}{\partial \mathbf{C}_k^{\mathrm{e}}}\mathbf{F}_k^{\mathrm{v}\text{-}\mathrm{T}}. \tag{2.65}$$

The Cauchy stress is further evaluated as

$$\boldsymbol{\sigma} = \frac{1}{J}\mathbf{F}\mathbf{S}\mathbf{F}^{\mathrm{T}}. \tag{2.66}$$

The remaining entropy inequality is written as

$$\sum_{k}^{N} 2\mathbf{C}_k^{\mathrm{e}}\frac{\partial \Psi_k^{\mathrm{neq}}}{\partial \mathbf{C}_k^{\mathrm{e}}} : \mathbf{D}_k^{\mathrm{v}} \geq 0. \tag{2.67}$$

Equation (2.67) can be satisfied through adopting the following evolution equation for $\mathbf{D}_k^{\mathrm{v}}$ as

$$\mathbf{D}_k^{\mathrm{v}} = \frac{\mathbf{M}_k}{2\eta_k}, \quad \mathbf{M}_k = 2\mathbf{C}_k^{\mathrm{e}}\frac{\partial \Psi_k^{\mathrm{neq}}}{\partial \mathbf{C}_k^{\mathrm{e}}}, \tag{2.68}$$

where $\mathbf{M}_k$ is the Mandel stress, η_k is the characteristic viscosity of the kth relaxation process. Similarly, we can define the relaxation time $\tau_k = \eta_k / G_k^{\mathrm{neq}}$.

We adopt the neo-Hookean model for both the equilibrium and nonequilibrium responses. The free energy can be expressed as

$$\Psi = \frac{G^{\mathrm{eq}}}{2}(\bar{I}_1 - 3) + \sum_{k}^{N} \frac{1}{2}G_k^{\mathrm{neq}}(\bar{I}_k^e - 3) + \frac{\kappa}{4}\left(J^2 - 2\ln(J) - 1\right), \tag{2.69}$$

where G^{eq} is the equilibrium shear modulus, G_k^{neq} is the shear modulus of kth nonequilibrium part, and κ is the bulk modulus, $\bar{I}_1 = \mathrm{tr}(\bar{\mathbf{C}})$, and $\bar{I}_k^e = \mathrm{tr}(\bar{\mathbf{C}}_k^{\mathrm{e}})$.

Substituting Eq. (2.69) into Eqs. (2.65) and (2.66) leads to the following form for the Cauchy stress:

$$\boldsymbol{\sigma} = \underbrace{\frac{1}{J}G^{\mathrm{eq}}(\bar{\mathbf{b}} - \frac{1}{3}\bar{I}_1\mathbf{I})}_{\sigma^{\mathrm{eq}}} + \sum_{k}^{N} \underbrace{\frac{1}{J}G_k^{\mathrm{neq}}\left(\bar{\mathbf{b}}_k^{\mathrm{e}} - \frac{1}{3}\bar{I}_k^e\mathbf{I}\right)}_{\sigma_k^{\mathrm{neq}}} + \underbrace{\frac{1}{2J}\kappa\left(J^2 - 1\right)\mathbf{I}}_{\mathbf{p}}. \tag{2.70}$$

The Mandel stress can be obtained as

$$\mathbf{M}_k = G_k^{\mathrm{neq}}(\bar{\mathbf{C}}_k^{\mathrm{e}} - \frac{1}{3}\bar{I}_k^e\mathbf{I}). \tag{2.71}$$

As discussed in Reese and Govindjee [47], the above model can be linearized to the small strain viscoelastic model.

2.3.2.2 Diffusion-based Transient Network Model

Linder et al. [48] developed a physical-based constitutive model, assuming that the viscous stress arises from the transient sub-network formed by temporary entanglements. The proposed model is based on micromechanically motivated diffusion processes of polymer chains described within the framework of Brownian motion. The free energy is also assumed to consist of an equilibrium part and viscous parts, while viscous parts are assumed to be a function of the internal variables $\mathbf{A}_k^{\mathrm{v}}$, represented as

$$\Psi = \Psi^{\mathrm{eq}}(\mathbf{C}) + \sum_k^N \Psi_k^{\mathrm{neq}}(\mathbf{C}, \mathbf{A}_k^{\mathrm{v}}). \tag{2.72}$$

The rate of Helmholtz free energy density is then expressed as

$$\dot{\Psi} = 2\frac{\partial \Psi^{\mathrm{eq}}}{\partial \mathbf{C}} : \frac{1}{2}\dot{\mathbf{C}} + 2\sum_k^N \frac{\partial \Psi_k^{\mathrm{neq}}}{\partial \mathbf{C}} : \frac{1}{2}\dot{\mathbf{C}} + \sum_k^N \frac{\partial \Psi_k^{\mathrm{neq}}}{\partial \mathbf{A}_k^{\mathrm{v}}} : \dot{\mathbf{A}}_k^{\mathrm{v}}. \tag{2.73}$$

Substituting the preceding equation into Eq. (2.63), we get

$$(\mathbf{S} - 2\frac{\partial \Psi^{\mathrm{eq}}}{\partial \mathbf{C}} - 2\sum_k^N \frac{\partial \Psi_k^{\mathrm{neq}}}{\partial \mathbf{C}}) : \frac{1}{2}\dot{\mathbf{C}} - \sum_k^N \frac{\partial \Psi_k^{\mathrm{neq}}}{\partial \mathbf{A}_k^{\mathrm{v}}} : \dot{\mathbf{A}}_k^{\mathrm{v}} \geq 0. \tag{2.74}$$

The equilibrium thermodynamic gives

$$\mathbf{S} = 2\frac{\partial \Psi^{\mathrm{eq}}}{\partial \mathbf{C}} + 2\sum_k^N \frac{\partial \Psi_k^{\mathrm{neq}}}{\partial \mathbf{C}}. \tag{2.75}$$

The entropy inequality can then be rewritten as

$$-\frac{\partial \Psi_k^{\mathrm{neq}}}{\partial \mathbf{A}_k^{\mathrm{v}}} : \dot{\mathbf{A}}_k^{\mathrm{v}} \geq 0. \tag{2.76}$$

For the equilibrium part, free energy is adopted the same form as in the multiplicative decomposition model as

$$\Psi^{\mathrm{eq}} = \frac{G^{\mathrm{eq}}}{2}(\bar{I}_1 - 3) + \frac{\kappa}{4}(J^2 - 2\ln(J) - 1). \tag{2.77}$$

The following form is adopted for the free energy in the viscous parts [48]:

$$\Psi_k^{\mathrm{neq}} = \frac{1}{2}G_k^{\mathrm{neq}}(\mathbf{A}_k^{\mathrm{v}} : \bar{\mathbf{C}} - 3 - \ln \det \mathbf{A}_k^{\mathrm{v}}). \tag{2.78}$$

Substituting Eqs. (2.77) and (2.78) into Eqs. (2.75) and (2.66) leads to the following form for the Cauchy stress:

$$\boldsymbol{\sigma} = \frac{1}{J}G^{\mathrm{eq}}(\bar{\mathbf{b}} - \frac{1}{3}\mathrm{tr}(\bar{\mathbf{b}})\mathbf{I}) + \sum_k^N \frac{1}{J}G_k^{\mathrm{neq}}\left[\bar{\mathbf{F}}\mathbf{A}_k^{\mathrm{v}}\bar{\mathbf{F}}^{\mathrm{T}} - \frac{1}{3}(\mathbf{A}_k^{\mathrm{v}} : \bar{\mathbf{C}})\mathbf{I}\right] + \frac{1}{2J}\kappa(J^2 - 1)\mathbf{I}. \tag{2.79}$$

The following evolution equation is adopted for the internal variable $\mathbf{A}_k^{\mathrm{v}}$ [48]:

$$\dot{\mathbf{A}}_k^{\mathrm{v}} = \frac{\bar{\mathbf{C}}^{-1} - \mathbf{A}_k^{\mathrm{v}}}{\tau_k}, \tag{2.80}$$

where τ_k is the relaxation time of the kth relaxation process. The symmetric tensors $\mathbf{A}_k^{\mathrm{v}}$ are initially set to being the identity tensor.

Remark. Vernerey et al. [49] have developed a statistically based model for transient networks. If assuming both the kinetic coefficient of chain breaking and reattaching is a constant, independent of the stretch/force, the model reduces to a viscoelastic model with the same form as Eq. (2.80).

Combining Eqs. (2.76), (2.78), and (2.80) yields

$$(\bar{\mathbf{C}} - \mathbf{A}_k^{\mathrm{v}-1}) : (\bar{\mathbf{C}}^{-1} - \mathbf{A}_k^{\mathrm{v}}) \leq 0. \tag{2.81}$$

Linder et al. [48] conducted a detailed proof to demonstrate the validity of this equation, which is also presented as follows:

Consider a polar decomposition of the two symmetric tensors $\bar{\mathbf{C}}$ and $\mathbf{A}_k^{\mathrm{v}}$ as well as their inverses in the form

$$\begin{aligned} \bar{\mathbf{C}} &= \sum_{i=1}^{3} \lambda_i \mathbf{u}_i \otimes \tilde{\mathbf{u}}_i, & \mathbf{A}_k^{\mathrm{v}} &= \sum_{j=1}^{3} \mu_j \mathbf{v}_j \otimes \tilde{\mathbf{v}}_j, \\ \bar{\mathbf{C}}^{-1} &= \sum_{i=1}^{3} \lambda_i^{-1} \tilde{\mathbf{u}}_i \otimes \mathbf{u}_i, & \mathbf{A}_k^{\mathrm{v}-1} &= \sum_{j=1}^{3} \mu_j^{-1} \tilde{\mathbf{v}}_j \otimes \mathbf{v}_j, \end{aligned} \tag{2.82}$$

where $\lambda_i > 0, \mu_j > 0$ are the positive eigenvalues and $\{\mathbf{u}_i, \tilde{\mathbf{u}}_i\}, \{\mathbf{v}_j, \tilde{\mathbf{v}}_j\}$ are the orthogonal eigenvectors of $\bar{\mathbf{C}}$ and $\mathbf{A}^{\mathrm{v}}$, respectively.

Since $\sum_i \mathbf{u}_i \otimes \tilde{\mathbf{u}}_i = \sum_j \mathbf{v}_j \otimes \tilde{\mathbf{v}}_j = \mathbf{I}$ and $\mathbf{I} : \mathbf{I} = 3$, the left-hand side of Eq. (2.81) becomes

$$(\bar{\mathbf{C}} - \mathbf{A}^{\mathrm{v}-1}) : (\bar{\mathbf{C}}^{-1} - \mathbf{A}^{\mathrm{v}}) = 6 - \sum_{i,j} [\lambda_i \mu_j + \lambda_i^{-1} \mu_j^{-1}] (\mathbf{u}_i \otimes \tilde{\mathbf{u}}_i) : (\mathbf{v}_j \otimes \tilde{\mathbf{v}}_j). \tag{2.83}$$

Considering $\alpha + \alpha^{-1} \geq 2$ for $\alpha > 0$, this leads to $\lambda_i \mu_j + \lambda_i^{-1} \mu_j^{-1} \geq 2$ and $(\mathbf{u}_i \otimes \tilde{\mathbf{u}}_i) : (\mathbf{v}_j \otimes \tilde{\mathbf{v}}_j) = (\mathbf{u}_i \cdot \mathbf{v}_j)(\tilde{\mathbf{u}}_i \cdot \tilde{\mathbf{v}}_j) = (\mathbf{u}_i \cdot \mathbf{v}_j)^2 \geq 0$ and finally

$$(\bar{\mathbf{C}} - \mathbf{A}^{\mathrm{v}-1}) : (\bar{\mathbf{C}}^{-1} - \mathbf{A}^{\mathrm{v}}) \leq 6 - 2 \cdot \mathbf{I} : \mathbf{I} = 0. \tag{2.84}$$

2.3.2.3 Viscoelastic Model Based on the Second Piola–Kirchhoff Stress

Simo [50] formulated a finite deformation viscoelastic model based on the internal variables $\mathbf{Q}_k$, which has a physical meaning as the nonequilibrium second Piola–Kirchhoff stress. The total second Piola–Kirchhoff stress can be represented as

$$\mathbf{S} = \mathbf{S}^{\mathrm{eq}} + \sum_{k}^{N} \mathbf{Q}_k, \tag{2.85}$$

where $\mathbf{S}^{\mathrm{eq}}$ is the equilibrium second Piola–Kirchhoff stress.

Define the tensor $\mathbf{S}_k^0$ as the isochoric second Piola–Kirchhoff stress in no relaxation state as

$$\mathbf{S}_k^0 = 2 \frac{\partial \Psi_k^0}{\partial \mathbf{C}}, \tag{2.86}$$

where Ψ_k^0 is the free energy density when no relaxation occurs.

The evolution equation of the $\mathbf{Q}_k$ is given as

$$\dot{\mathbf{Q}}_k + \frac{\mathbf{Q}_k}{\tau_k} = \dot{\mathbf{S}}_k^0, \tag{2.87}$$

where τ_k is the relaxation time.

To complete the model, we use the same form of the free energy for the equilibrium part as in previous finite deformation viscoelastic models, which give

$$\mathbf{S}^{\mathrm{eq}} = G^{\mathrm{eq}} J^{-2/3} \left(\mathbf{I} - \frac{1}{3} \mathrm{tr}(\mathbf{C}) \mathbf{C}^{-1} \right) + \frac{1}{2} \kappa \left(J^2 - 1 \right) \mathbf{C}^{-1}. \tag{2.88}$$

We adopt the neo-Hookean model for Ψ_k^0 as

$$\Psi_k^0 = \frac{G_k^{\mathrm{neq}}}{2} \left(\bar{I}_1 - 3 \right). \tag{2.89}$$

Combining Eqs. (2.86) and (2.89) gives

$$\mathbf{S}_k^0 = G_k^{\mathrm{neq}} J^{-2/3} \left(\mathbf{I} - \frac{1}{3} \mathrm{tr}(\mathbf{C}) \mathbf{C}^{-1} \right). \tag{2.90}$$

The Cauchy stress can then be evaluated through Eq. (2.66).

2.3.2.4 Viscoelastic Model Based on the Cauchy Stress

Since the viscoelastic model can be constructed based on the second Piola–Kirchhoff stress in the reference configuration, it is naturally to develop a corresponding viscoelastic model based on the Cauchy stress [16, 51, 52]. These models are formulated either in the integral form or differential form [52]. Here, to be consistent with previous parts, we only introduce the differential form.

The total stress can be represented as

$$\boldsymbol{\sigma} = \boldsymbol{\sigma}^{\mathrm{eq}} + \sum_k^N \boldsymbol{\sigma}_k^{\mathrm{neq}}, \tag{2.91}$$

where $\boldsymbol{\sigma}^{\mathrm{eq}}$ is the equilibrium Cauchy stress and $\boldsymbol{\sigma}_k^{\mathrm{neq}}$ is the nonequilibrium Cauchy stress of the kth relaxation process.

The viscoelastic stress can be determined using the following equation:

$$\dot{\boldsymbol{\sigma}}_k^{\mathrm{neq}} + \frac{\boldsymbol{\sigma}_k^{\mathrm{neq}}}{\tau_k} = \dot{\boldsymbol{\sigma}}_k^0, \tag{2.92}$$

where $\boldsymbol{\sigma}_k^0$ is the viscous stress of the kth relaxation process when no relaxation occurs.

Using the neo-Hookean model for both the equilibrium and viscous stresses, the explicit form of $\boldsymbol{\sigma}^{\mathrm{eq}}$ and $\boldsymbol{\sigma}_k^0$ can be represented as

$$\boldsymbol{\sigma}^{\mathrm{eq}} = \frac{1}{J} G^{\mathrm{eq}} \left(\bar{\mathbf{b}} - \frac{1}{3} \mathrm{tr}(\bar{\mathbf{b}}) \mathbf{I} \right) + \frac{1}{2J} \kappa \left(J^2 - 1 \right) \mathbf{I}, \tag{2.93a}$$

$$\boldsymbol{\sigma}_k^0 = \frac{1}{J} G_k^{\mathrm{neq}} \left(\bar{\mathbf{b}} - \frac{1}{3} \mathrm{tr}(\bar{\mathbf{b}}) \mathbf{I} \right). \tag{2.93b}$$

Remark. For the two viscoelastic models based on the stress approach, we have not provided a proof on the thermodynamic consistent of these models. In fact, as pointed out by

Govindjee and Reese [53], these model cannot be conclusively proven to satisfy the second law of thermodynamics. In the small strain condition, the above two models can be reduced to the classic viscoelastic models with the strict proof of entropy inequality as shown in Christensen [51].

2.3.3 Results

We have presented four finite deformation viscoelastic models. To be convenient, we name the above models as Model 1–4 as the order of the above presentation. It is necessary to compare the performance of these models. Since the main interest is regarding the viscoelastic response, the equilibrium shear modulus is set as 0, while only one relaxation process is adopted for the viscous process. The nonequilibrium shear modulus G^{neq} is chosen as 1/3 MPa, which gives a Young's modulus of 1 MPa in the small strain loading condition. The relaxation time τ is specified as 100 s.

Here, we show two different loading cases. For the first condition, the step stretch $\lambda_1(t) = \lambda_0$ is applied at time 0 and then held. The stress is measured as a function of time. It is straightforward to show that for Model 2, 3 and 4, the stress response can be analytically obtained as

$$\sigma(t) = G^{\text{neq}}\left(\lambda_0^2 - \frac{1}{\lambda_0}\right)\exp\left(-\frac{t}{\tau}\right). \tag{2.94}$$

The stress response of Model 1 has to be evaluated numerically. As shown in Figure 2.6, when the stretch is small ($\lambda_0 = 1.1$), all the four models have almost identical stress relaxation response, which is almost the same as the small strain viscoelastic model. When the stretch ratio is large, for example, $\lambda_0 = 2.0$, the multiplicative decomposition model (Model 1) exhibits a faster relaxation process compared with the other three models.

We further evaluate the performance of these models through a loading–unloading cycle at a constant loading rate, namely, $\dot{\lambda}_1 = \text{constant}$. We first consider a maximum stretch of 1.1, which can be characterized as the small strain regime. As shown in Figure 2.7, the response transits from elastic to viscoelastic with a decreasing strain rate. At the lowest strain rate (0.0001/s), a clear stress plateau is observed similar to a fluid response. The four models yield almost the same stress response, especially when the strain rate is higher.

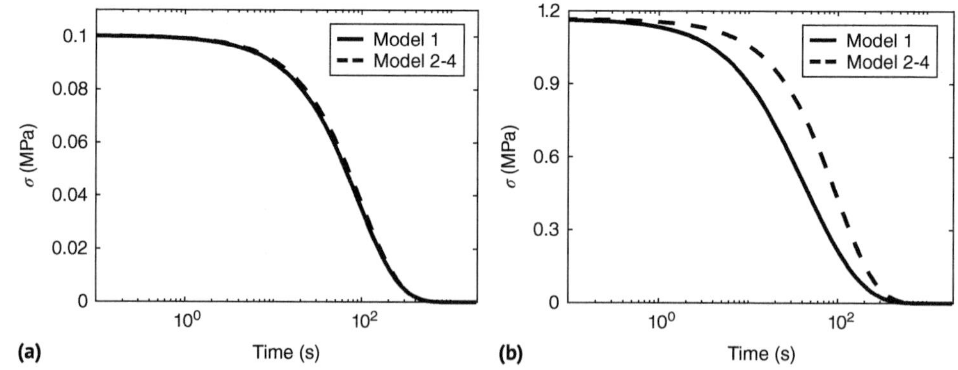

Figure 2.6 Stress relaxation of the finite deformation viscoelastic models: (a) $\lambda_0 = 1.1$ and (b) $\lambda_0 = 2.0$.

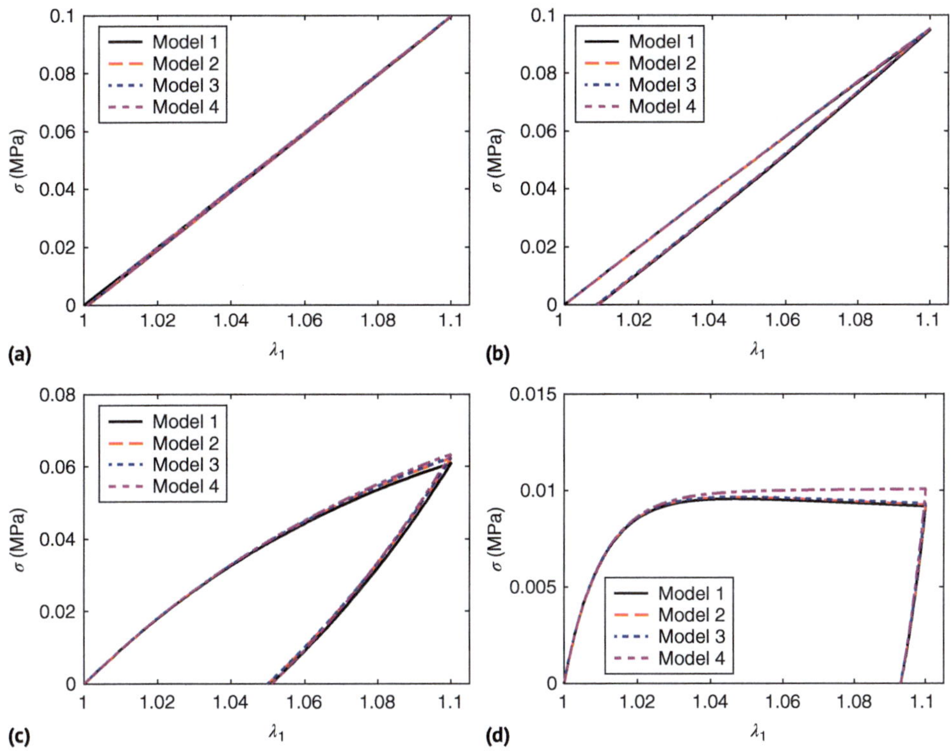

Figure 2.7 Stress response of different finite deformation viscoelastic models with the maximum stretch 1.1 loaded at (a) 0.1/s, (b) 0.01/s, (c) 0.001/s and (d) 0.0001/s.

Considerable difference can be observed for the four viscoelastic models when the deformation is large, as shown in Figure 2.8. At the strain rate of 0.1/s, negligible viscoelasticity can be observed, though more relaxation can still occur for Model 1. With decreasing the strain rate, more relaxation can be observed. In all cases, more relaxation occurs in Model 1, while the least relaxation occurs for Model 4. It is also observed that the two models based on the strain approach (Model 1 and Model 2) overlap when the strain rate is lowest. Since for most of the viscoelastic models, the relaxation spectrum ($G_k^{\mathrm{neq}} - \tau_k$) can be obtained through small strain tests, such as dynamic mechanical analysis and stress relaxation tests. The model parameters will be identical for all the models. Employing different models can then yield varied predictions. Thus, it will be of interest to fully evaluate these models in predicting the mechanical performance of various soft materials.

2.3.4 Finite Element Simulation

We have presented several hyperelastic and viscoelastic models in this chapter. Some of these models have been integrated into commercial finite element software, such as ABAQUS and COMSOL Multiphysics. Since the book focuses on the multi-field coupled behaviors in soft materials, we employ the COMSOL Multiphysics to implement these models. In COMSOL, the newly developed hyperelastic models can be implemented through providing the explicit form of the free energy function, which is straightforward.

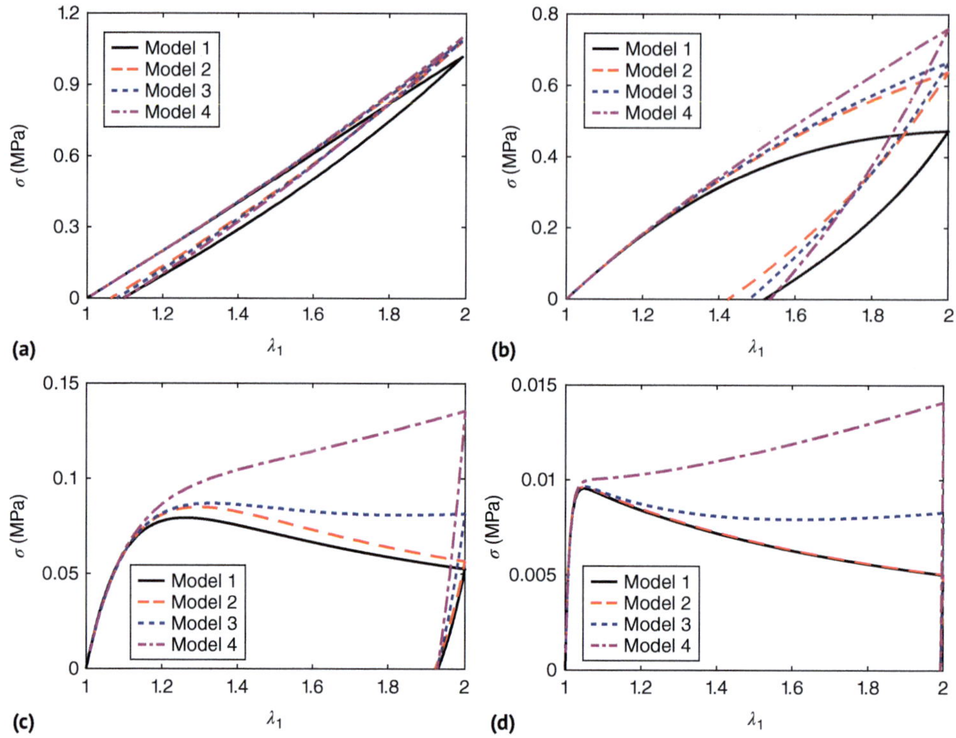

Figure 2.8 Stress response of different finite deformation viscoelastic models with the maximum stretch 2.0 loaded at (a) 0.1/s, (b) 0.01/s, (c) 0.001/s, and (d) 0.0001/s.

The viscoelastic model based on the second Piola–Kirchhoff stress has already been built into COMSOL. It is relatively complicated to implement the multiplicative decomposition model into COMSOL, while this model has been implemented into several open source finite element suites, such as FEAP, Tahoe using the algorithm proposed in Reese and Govindjee [47].

We have implemented the other two viscoelastic models into COMSOL. We have used the Hyperelastic Material in Solid Mechanics Module. The viscoelastic stress is added as an external stress, which is represented as the second Piola–Kirchhoff stress. For the diffusion-based transient network model, the evolution of the internal variable $\mathbf{A}_k^{\mathrm{v}}$ (Eq. 2.80) can be solved using the coefficient form PDE interfaces in COMSOL. The second Piola–Kirchhoff stress can then be obtained through combining Eqs. (2.21) and (2.78).

The implementation of the viscoelastic model based on the Cauchy stress follows the similar way. The nonequilibrium Cauchy stress needs to be pulled back to the second Piola–Kirchhoff stress. In order to simplify the numerical procedures, we define $\mathbf{q}_k = \boldsymbol{\sigma}_k^0 - \boldsymbol{\sigma}_k^{\mathrm{neq}}$. Equation (2.92) can then be rewritten as

$$\dot{\mathbf{q}}_k + \frac{\mathbf{q}_k}{\tau_k} = \frac{\boldsymbol{\sigma}_k^0}{\tau_k}. \tag{2.95}$$

The preceding equation can be solved using the coefficient form PDE interfaces in COMSOL, which is further used to obtain the nonequilibrium Cauchy stress.

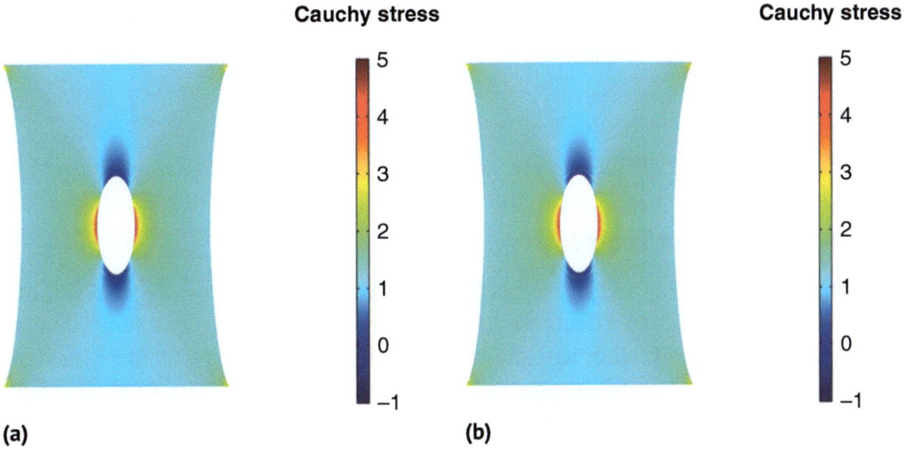

Figure 2.9 The stress contours obtained using: (a) the diffusion-based transient network model and (b) the viscoelastic model based on the Cauchy stress.

We show one numerical example of the performance of the above two models in the plane stress condition. A hole with a radius of 0.1 mm is cut in the middle of a specimen of 1 mm × 1 mm. The bottom of the specimen is then fixed, while the top edge is pulled in the vertical direction at a constant engineering strain rate of 0.03/s. For the hyperelastic part, a neo-Hookean model is used with $G^{\mathrm{eq}} = 0.5$ MPa. Only one relaxation process is adopted with $G^{\mathrm{neq}} = 0.5$ MPa and $\tau = 150$ s. The stress contours at the time 14.5 s is shown in Figure 2.9, which shows a very similar response for the two models.

2.4 Damage Models

In this section, we first present several classic damage models. We then show two extensions on these models to describe stress softening behaviors in tough elastomers and gels.

2.4.1 Continuum Damage Model

For the continuum damage model, it is assumed that the deformation induces a progressively reduction of the free energy density. Thus, the free energy density can be represented as

$$\Psi = (1 - \zeta)\Psi_0, \tag{2.96}$$

where Ψ_0 is the strain energy function of the undamaged material. The variable $0 \le \zeta \le 1$ represents the damage level.

The thermodynamics law requires that $\dot{\eta} \ge 0$, while the hyperelastic models can be adopted for the free energy function of the undamaged condition. Miehe [54] adopted the following form for the damage variable:

$$\zeta = \zeta_\infty \left[1 - \exp\left(-\frac{\alpha}{l}\right)\right], \tag{2.97}$$

where ζ_∞ is the maximum damage level and l is the damage saturation parameter.

It is further assumed that

$$\alpha(t) = \max \Psi_0(s), \tag{2.98}$$

where s denotes the whole loading history.

2.4.2 Network Alteration Theory

The network alteration theory [31] assumes that breakage of links is the physical origin of the Mullins effect. This leads to a decrease in the number of chain density and an increase in the chain segments. It is further assumed that the total chain segments are conserved. Thus, the network alteration theory can naturally be related with the model developed based on the Langevin statistics. Use the eight-chain model in Section 2.2.2.5 as an example. The free energy density is the same as Eq. (2.39). The shear modulus and the number of chain segments can be represented as

$$G = \frac{G_0}{f_0}, \quad N = N_0 f_0, \tag{2.99}$$

where G_0 is the initial modulus of undamaged network, N_0 is the initial number of chain segment, and $f_0 \geq 1$ is the damage function.

Xiao et al. 34 proposed the following function for f_0 to describe the damage behavior of double network hydrogels as

$$f(I_1^{\max}) = \exp(a_1(I_1^{\max} - 3)), \tag{2.100}$$

where $I_1^{\max}$ represents the maximum first invariant that the materials have been experienced over all the loading history, and a_1 is a pre-factor for the sensitivity of damage level to the applied deformation. This is an isotropic damage model.

2.4.3 Progressively Damage Model

The progressively damage model assumes a broad distribution of polymer chain lengths. As deformation, the short chains break first, while the long chains survive to withstand the loads. Here we use the eight-chain model to represent the free energy function of each individual chain. Thus, the total free energy density can be represented as [13]

$$\Psi = G \sum_i \phi_i N_i \left(\frac{\bar{\lambda}}{\sqrt{N_i}} \beta_i + \ln \frac{\beta_i}{\sinh \beta_i} \right) + \Psi_0, \quad \bar{\lambda} = \sqrt{\frac{I_1}{3}}, \quad \beta_i = \mathcal{L}^{-1}\left(\frac{\bar{\lambda}}{\sqrt{N_i}} \right), \tag{2.101}$$

Here, G represents the shear modulus of the total network, and ϕ_i is the number fraction of polymer chain with chain segments N_i.

It is typically assumed chain breaks when reach a certain critical stress. This can be further rewritten as $\phi_i = 0$, when $\frac{\bar{\lambda}}{\sqrt{N_i}} \geq \lambda_{\text{cri}}$.

2.4.4 Extended Network Alteration Theory

The network alteration theory presented in Section 2.4.2 is inherently an isotropic damage model. Here we first extend this model to the anisotropic damage condition using the affine assumption.

The free energy density of the total system can be written in the following integration form as

$$\Psi = \frac{1}{|S|}\int_S nkTN\left(\frac{\lambda\beta}{\sqrt{N}} + \ln\frac{\beta}{\sinh\beta}\right)dS, \qquad \beta(\mathbf{r}) = \mathcal{L}^{-1}\left(\frac{\lambda}{\sqrt{N}}\right), \tag{2.102}$$

where $|S|$ is its surface, $\lambda = \lambda(\mathbf{r})$, $n = n(\mathbf{r})$ and $N = N(\mathbf{r})$ all depend on the direction of chains.

We then rewrite the preceding equation through defining the initial shear modulus $G_0 = n_0 kT$ and the normalized number density of chains $q(\mathbf{r}) = n(\mathbf{r})/n_0$:

$$\Psi = G_0\frac{1}{|S|}\int_S qN\left(\frac{\lambda\beta}{\sqrt{N}} + \ln\frac{\beta}{\sinh\beta}\right)dS. \tag{2.103}$$

Using the affine assumption, the stretch ratio in the $\mathbf{r}$ direction can be represented as $\lambda = \sqrt{\mathbf{r}\cdot\mathbf{Cr}}$.

We can further adopt the discrete method in Bažant and Oh [55] to calculate the integration on a microsphere. The chain direction vectors $\mathbf{r}_j = \alpha_j\mathbf{e}_1 + \beta_j\mathbf{e}_2 + \gamma_j\mathbf{e}_3$ and weights ω_j can be found in Bažant and Oh [55]. The free energy density can be rewritten in a discrete manner as

$$\Psi = G_0\sum_j q_j N_j\left(\frac{\lambda_j\beta_j}{\sqrt{N_j}} + \ln\frac{\beta_j}{\sinh\beta_j}\right)\omega_j, \qquad \beta_j = \mathcal{L}^{-1}\left(\frac{\lambda_j}{\sqrt{N_j}}\right), \qquad \lambda_j = \sqrt{\lambda_1^2\alpha_j^2 + \lambda_2^2\beta_j^2 + \lambda_3^2\gamma_j^2}. \tag{2.104}$$

The corresponding Cauchy stress in the principal direction can then be represented as

$$\sigma_1 = G_0\sum_j q_j\sqrt{N_j}\beta_j\frac{\lambda_1^2\alpha_j^2\omega_j}{\lambda_j} + p, \qquad \sigma_2 = G_0\sum_j q_j\sqrt{N_j}\beta_j\frac{\lambda_2^2\beta_j^2\omega_j}{\lambda_j} + p, \qquad \sigma_3 = G_0\sum_j q_j\sqrt{N_j}\beta_j\frac{\lambda_3^2\gamma_j^2\omega_j}{\lambda_j} + p, \tag{2.105}$$

where p is the hydrostatic pressure.

The following evolution equation is adopted to represent the damage in each individual direction:

$$q_j = \frac{1}{1 + a_2(\lambda_j^{\max} - 1)^2}, \quad N_j = N_0\left(1 + a_2(\lambda_j^{\max} - 1)^2\right), \tag{2.106}$$

where a_2 is a pre-factor of the damage function and $\lambda_j^{\max}$ is the maximum stretch ratio that the chains have experienced over all the deformation history in the jth direction.

Equations (2.105) and (2.106) consist of the governing equations of the affine microsphere damage model.

We then show another extension on the isotropic damage model to the non-affine microsphere damage model. The following kinetic constraint is adopted:

$$\frac{1}{|S|}\int_S \boldsymbol{\lambda} \otimes \mathbf{r} dS - \frac{\mathbf{F}}{3} = 0, \tag{2.107}$$

where $\boldsymbol{\lambda}$ is the stretch vector of the initial direction $\mathbf{r}$. The physical meaning of Eq. (2.107) represents that the macroscopic strain is an average of the microscopic strain.

The free energy density function of the non-affine microsphere damage model is the same as that of the affine model. Thus, the kinematic constraint can be further incorporated into the free energy density, which gives

$$L(\boldsymbol{\lambda}, \boldsymbol{\Pi}) = \frac{1}{|S|}\int_S nkTN\left(\frac{\lambda\beta}{\sqrt{N}} + \ln\frac{\beta}{\sinh\beta}\right) dS - \boldsymbol{\Pi} : \left(\frac{1}{|S|}\int_S \boldsymbol{\lambda} \otimes \mathbf{r} dS - \frac{\mathbf{F}}{3}\right), \tag{2.108}$$

where the second-order tensor $\boldsymbol{\Pi}$ is the Lagrangian multiplier.

Equation (2.108) can be further rewritten in the discrete form as

$$L(\boldsymbol{\lambda}_j, \boldsymbol{\Pi}) = \sum_j n_j kTN_j\left(\frac{\lambda_j\beta_j}{\sqrt{N_j}} + \ln\frac{\beta_j}{\sinh\beta_j}\right)\omega_j - \boldsymbol{\Pi} : \left(\sum_j \boldsymbol{\lambda}_j \otimes \mathbf{r}_j\omega_j - \frac{\mathbf{F}}{3}\right), \tag{2.109}$$

where $\beta_j = \mathcal{L}^{-1}(\lambda_j/\sqrt{N_j})$, and λ_j is the corresponding stretch ratio in the jth direction.

The stationary condition can be satisfied using the following relation:

$$\frac{\partial L}{\partial \boldsymbol{\lambda}_j} = 0, \tag{2.110a}$$

$$\frac{\partial L}{\partial \boldsymbol{\Pi}} = 0. \tag{2.110b}$$

Substituting Eq. (2.109) into Eq. (2.110) gives

$$n_j kT\sqrt{N_j}\beta_j\frac{\boldsymbol{\lambda}_j}{\lambda_j} = \boldsymbol{\Pi}\mathbf{r}_j, \tag{2.111a}$$

$$\sum_j \boldsymbol{\lambda}_j \otimes \mathbf{r}_j\omega_j - \frac{\mathbf{F}}{3} = 0. \tag{2.111b}$$

Equation (2.111a) provides the relationship between the microscopic stretch ratio of polymer chains with the Lagrangian multiplier as

$$\lambda_j = \sqrt{N_j}\mathcal{L}\left(\frac{||\boldsymbol{\Pi}\mathbf{r}_j||}{n_j kT\sqrt{N_j}}\right), \tag{2.112}$$

where $||\mathbf{\Pi}\mathbf{r}_j||$ is the norm of the vector $\mathbf{\Pi}\mathbf{r}_j$.

Combining Eqs. (2.111a), (2.111b), and (2.112) gives the following kinematic constraint:

$$\boldsymbol{\Pi}\cdot\sum_j \sqrt{N_j}\mathcal{L}\left(\frac{||\boldsymbol{\Pi}\mathbf{r}_j||}{n_j kT\sqrt{N_j}}\right)\frac{\mathbf{r}_j\otimes\mathbf{r}_j}{||\boldsymbol{\Pi}\mathbf{r}_j||}\omega_j = \frac{\mathbf{F}}{3}. \tag{2.113}$$

Given the current state of the number density of chains n_j, the number of chain segments N_j and the deformation gradient $\mathbf{F}$, the Lagrangian multiplier $\mathbf{\Pi}$ can be calculated based on Eq. (2.113).

The Cauchy stress is directly related with the Lagrangian multiplier as [34]

$$\boldsymbol{\sigma} = \frac{\mathbf{\Pi}\mathbf{F}^{\mathrm{T}}}{3}. \tag{2.114}$$

The damage function in each direction is introduced through the evolution equations of the normalized number density of chains $q_j = n_j/n_0$ and the number of chain segments N_j as

$$\begin{aligned} q_j &= \frac{1}{1 + a_3\left(\beta_j^{\max} - \beta_j^{\mathrm{ini}}\right)^2}, \\ N_j &= N_0\left(1 + a_3\left(\beta_j^{\max} - \beta_j^{\mathrm{ini}}\right)^2\right), \end{aligned} \tag{2.115}$$

where a_3 is a pre-factor of the damage function, $\beta_j^{\mathrm{ini}} = \mathcal{L}^{-1}\left(1/\sqrt{N_0}\right)$, and $\beta_j^{\max}$ is the maximum β_j over all the deformation history in the jth direction. Since the chain force of the cross-linked network is proportional to β_j, the above evolution equation indicates that the damage level depends on the maximum tensile force that the cross-linked chains have experienced in the whole deformation history.

Given the initial model parameters and loading history, the stress can then be obtained through combining Eqs. (2.113)–(2.115).

2.4.5 Extended Progressively Damage Model

It is found that the progressive damage model is not able to describe the mechanical behaviors of multiple network elastomers. One of the main reasons is that the first network in multiple network elastomers is prestreched, while the existence of the matrix can allow that the first networks reach close to the stretch limit. The classic hyperelastic model assumes that the free energy change is only caused by the entropic change. This cannot describe the behavior of single chain in multiple network elastomers.

Mao et al. [56] proposed a new single chain model, which assumed that the total deformation is composed of the chain stretch λ^c and the bond deformation λ^b as $\lambda = \lambda^b\lambda^c$. The free energy of the chain stretch contains a change in entropy and an enthalpy change caused by the bond deformation. Thus, the free energy density of a single chain is extended as

$$w_1(\lambda, \lambda^b) = kTN\left(\frac{\lambda}{\lambda^b\sqrt{N}}\beta + \ln\frac{\beta}{\sinh\beta}\right) + \frac{1}{2}NE_b(\ln\lambda^b)^2 + \text{constant},$$
$$\beta = \mathcal{L}^{-1}\left(\frac{\lambda}{\lambda^b\sqrt{N}}\right), \tag{2.116}$$

where E_b represents the stiffness of the bond.

Minimizing the free energy density of a single chain gives the value of λ^b as

$$\frac{\partial w_1(\lambda, \lambda^b)}{\partial\lambda^b} = 0, \tag{2.117}$$

which further yields

$$\frac{E_b}{kT}\lambda^b \ln\lambda^b = \frac{\lambda}{\sqrt{N}}\beta. \tag{2.118}$$

For the filler network, the total number of polymer chains is n_f and the number of polymer chains with chain segment N_j is n_j. Thus, the number fraction of polymer chains with the chain segment number N_j is defined as $P_j = n_j/n_\mathrm{f}$. The free energy density of the filler network is then written as

$$W_\mathrm{f}(\lambda, \lambda_j^b) = n_f\sum_j P_j\left[kTN_j\left(\frac{\lambda}{\lambda_j^b\sqrt{N_j}}\beta_j + \ln\frac{\beta_j}{\sinh\beta_j}\right) + \frac{1}{2}N_jE_b(\ln\lambda_j^b)^2\right] + \text{constant},$$
$$\beta_j = \mathcal{L}^{-1}\left(\frac{\lambda}{\lambda_j^b\sqrt{N_j}}\right), \tag{2.119}$$

where λ_j^b is the bond stretch of chain with chain segment number N_j.

It is further assumed the matrix can be represented by a hyperelastic model as

$$W_\mathrm{m}(\lambda) = n_m kTN_\mathrm{m}\left(\frac{\lambda}{\sqrt{N_\mathrm{m}}}\beta + \ln\frac{\beta}{\sinh\beta}\right) + \text{constant},$$
$$\beta = \mathcal{L}^{-1}\left(\frac{\lambda}{\sqrt{N_\mathrm{m}}}\right), \tag{2.120}$$

where n_m is the number of chains per unit volume and N_m is the number of chain segments in the matrix.

For a multiple network elastomer, each network is subjected to different levels of pre-stretch. The first network has a pre-strech ratio as λ_{s1}, while the ith network has a pre-strech ratio as λ_{si}. The corresponding volumetric ratio of each network can then be obtained by assuming the volume incompressible during mixing. Thus, the volumetric ratio of the first network is $\phi_1 = 1/\lambda_{s1}^3$. The corresponding volumetric ratio of the ith network is $\phi_i = 1/\lambda_{si}^3 - 1/\lambda_{s(i-1)}^3$.

The eight-chain model [13] is used to represent the equivalent stretch of chain. The equivalent stretch has contributions from pre-strech induced in the synthesized process and the mechanical deformation, and can be represented as $\tilde{\lambda}_i = \lambda_{si}\sqrt{I_1/3}$ with λ_{si} being the pre-stretch ratio and I_1 being the first invariant of the right Cauchy–Green deformation tensor $\mathbf{C} = \mathbf{F}^{\mathrm{T}}\mathbf{F}$.

The free energy density per reference volume can then be written as [41]

$$\Psi\left(\lambda, \lambda_{ij}^b\right) = \phi_1 n_\mathrm{f} \sum_j P_j \left[kTN_j\left(\frac{\tilde{\lambda}_1}{\lambda_j^b\sqrt{N_j}}\beta_j + \ln\frac{\beta_j}{\sinh\beta_j}\right) + \frac{1}{2}N_j E_b \left(\ln\lambda_j^b\right)^2\right]$$
$$+ \sum_{i=2} \phi_i n_m kTN_\mathrm{m}\left(\frac{\tilde{\lambda}_i}{\sqrt{N_\mathrm{m}}}\beta_{mi} + \ln\frac{\beta_{mi}}{\sinh\beta_{mi}}\right) + \text{constant},$$
$$\beta_j = \mathcal{L}^{-1}\left(\frac{\tilde{\lambda}_1}{\lambda_j^b\sqrt{N_j}}\right),$$
$$\beta_{mi} = \mathcal{L}^{-1}\left(\frac{\tilde{\lambda}_i}{\sqrt{N_\mathrm{m}}}\right), \tag{2.121}$$

The $\mu_\mathrm{f} = n_\mathrm{f}kT$ and $\mu_\mathrm{m} = n_\mathrm{m}kT$ are defined as the shear moduli of the filler network and matrix, respectively.

Using the incompressible condition, the stress can be represented as

$$\sigma = \phi_1 \sum_j P_j \frac{\mu_\mathrm{f}\sqrt{N_j}}{3}\frac{\lambda_{s1}^2}{\tilde{\lambda}_1}\frac{\beta_j}{\lambda_j^b}\mathbf{b} + \sum_{i=2}\phi_i \frac{\mu_\mathrm{m}\sqrt{N_m}}{3}\frac{\lambda_{si}^2}{\tilde{\lambda}_i}\beta_{mi}\mathbf{b} - p\mathbf{I}. \tag{2.122}$$

Equation (2.118) can be rewritten as to obtain the bond deformation:

$$\frac{E_b}{kT}\sqrt{N_j\lambda_j^b}\ln\lambda_j^b = \tilde{\lambda}_1\beta_j. \tag{2.123}$$

The damage criterion can be expressed as

$$P_j = 0, \text{when } \frac{\tilde{\lambda}_1^{max}}{\sqrt{N_j}} \geq \lambda_\mathrm{cri}, \tag{2.124}$$

where $\tilde{\lambda}_1^{max}$ is the maximum equivalent stretch ratio in the synthesized and loading history.

2.4.6 Results

Xiao et al. [34] applied the network alteration theory to describe the mechanical response of double network hydrogels subjected to uniaxial, pure shear (loaded in $\mathbf{e}_1$ direction) and equibiaxial loading conditions [57]. The experimental results show that a clear hysteresis loop can be observed for the loading–unloading cycles, while the reloading curve follows the unloading curve until reaching the previous maximum stretch. This indicates that double network hydrogels do not exhibit a viscoelastic response. As shown in Figures 2.10–2.12, all the three models based on network alteration theory can generally reproduce the main features of experimental results. However, the isotropic damage model overestimates the stress magnitude and the hysteresis area in $\mathbf{e}_2$ direction. In contrast, the affine damage model underestimates the stress magnitude and the hysteresis area in $\mathbf{e}_2$ direction. Compared with these two models, the non-affine damage model can well predict the stress response in loading conditions. It slightly overestimates the stress level in $\mathbf{e}_2$ direction when loaded in pure shear condition.

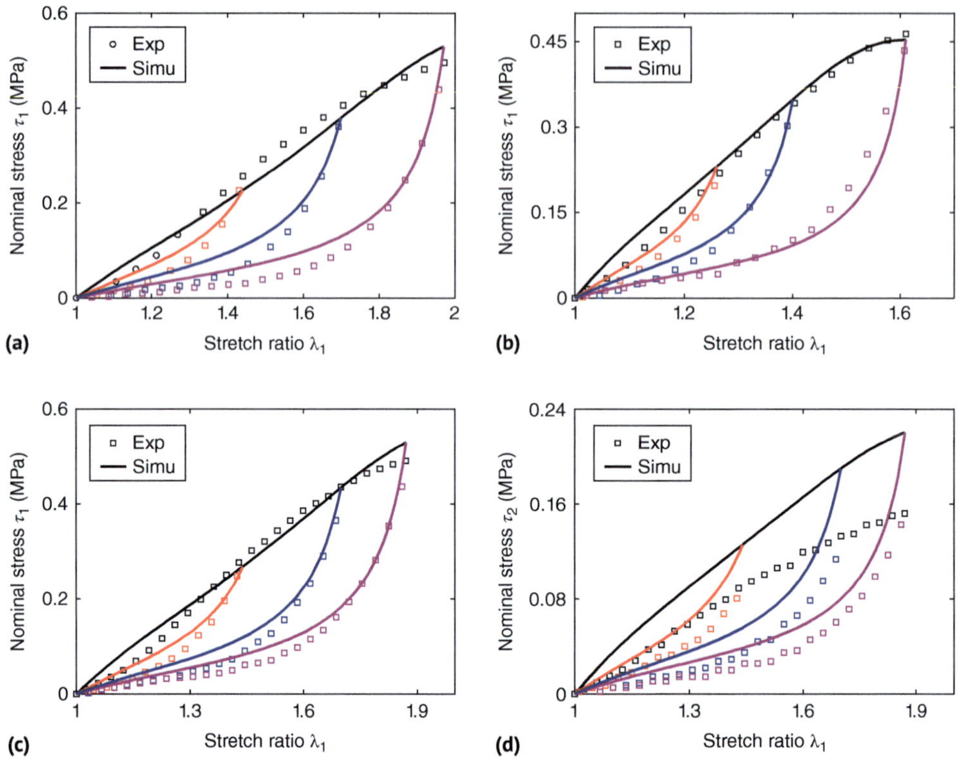

Figure 2.10 The performance of the isotropic damage model: (a) uniaxial tension, (b) equibiaxial tension, (c) pure shear (stress in $\mathbf{e}_1$ direction), and (d) pure shear (stress in $\mathbf{e}_2$ direction) with $G_0 = 0.0464$ MPa, $N_0 = 1.188$, and $a_1 = 1.45$.

The extended progressively damage model is further employed to describe the experimental data of multiple network elastomers. The model parameters are listed in Tables 2.1 and 2.2. The distribution of chain lengths is provided in Figure 2.13. The pre-stretch ratio is directly obtained from the experimental synthesis process, which is not a fitted parameter. The shear modulus μ_f is obtained through fitting to the single network elastomer. It is found that the modulus of the matrix has a minimal effect on the stress response. The number of chain segment is set as 100, which leads to almost neo-Hookean-type response for the matrix. The method proposed in Govindjee and Simo [58] is used to generate the distribution of polymer chains. The distribution of chain segments together with the parameters related with the extended single chain model is further fitted to the stress response of triple network elastomers with $\lambda_{s1} = 2.41$. For detailed information, refer to Xiao et al. [41].

Figure 2.14 evaluates the performance of the extended progressively damage model through comparison with the measured stress response of elastomers with different number of networks. The model can well predict the experimental results for all elastomers. Millereau [59] found that the stress response of single network elastomers is sensitive to

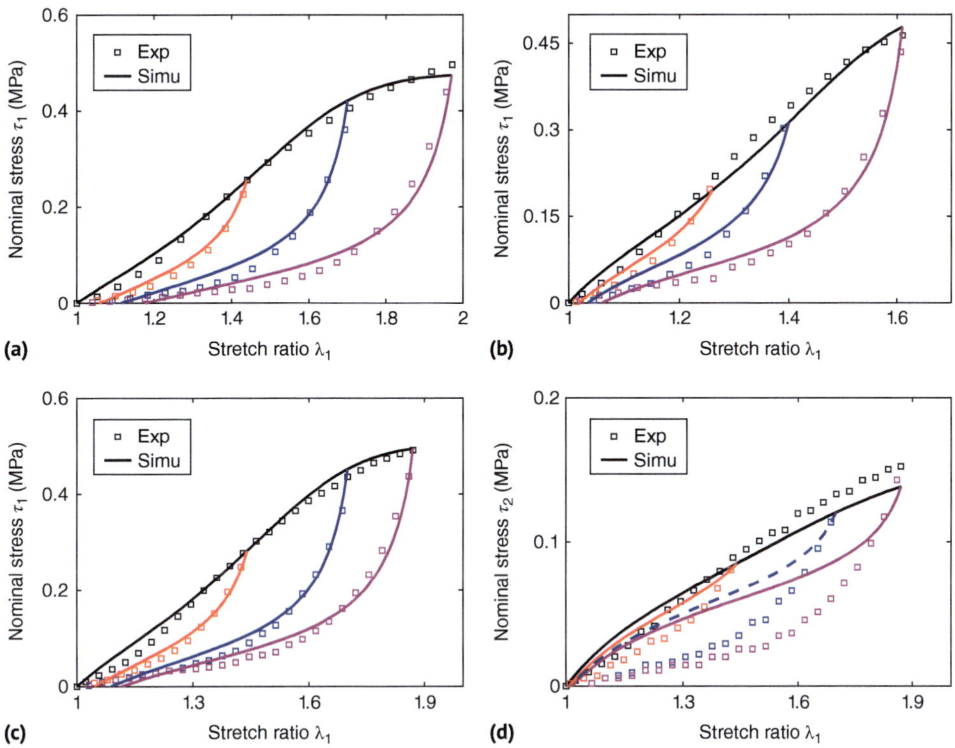

Figure 2.11 The performance of the affine damage model: (a) uniaxial tension, (b) equibiaxial tension, (c) pure shear (stress in $\mathbf{e}_1$ direction), and (d) pure shear (stress in $\mathbf{e}_2$ direction) with $G_0 = 0.063$ MPa, $N_0 = 1.75$, and $a_2 = 1.63$.

the synthesis conditions. As shown in Figure 2.14a, the ratio between the upper and low bound of stress can reach as high as 1.35. A median value of the modulus is chosen in the simulation for the filler network. Due to pre-stretch, a portion of the chains may already reach the stretch limit. Thus, the distribution of chain lengths greatly influences the stress response in multiple network elastomers. The model can well capture the initial smooth increase in stress and the subsequent significant hardening response for double and triple elastomer networks (Figure 2.14b, c). The predicted onset point of hardening response agrees well with the experimental observations. For quadruple network elastomers, due to a high initial pre-stretching ratio, the stress increases linearly with stretch until reaching a plateau, which is also fully captured by the model.

The model is further applied to simulate the stress response of a triple network elastomer in loading–unloading cycles. As shown in Figure 2.15, the elastomer with $\lambda_{s1} = 2.41$ shows the stress softening effect. A clear hysteresis loop can be observed. The model captures the main features of experimental data. However, it overestimates the stress in the initial stage. The Mullins effect is assumed to be caused by the progressive damage of chains. With increasing the maximum stretch ratio, more and more chains are ruptured.

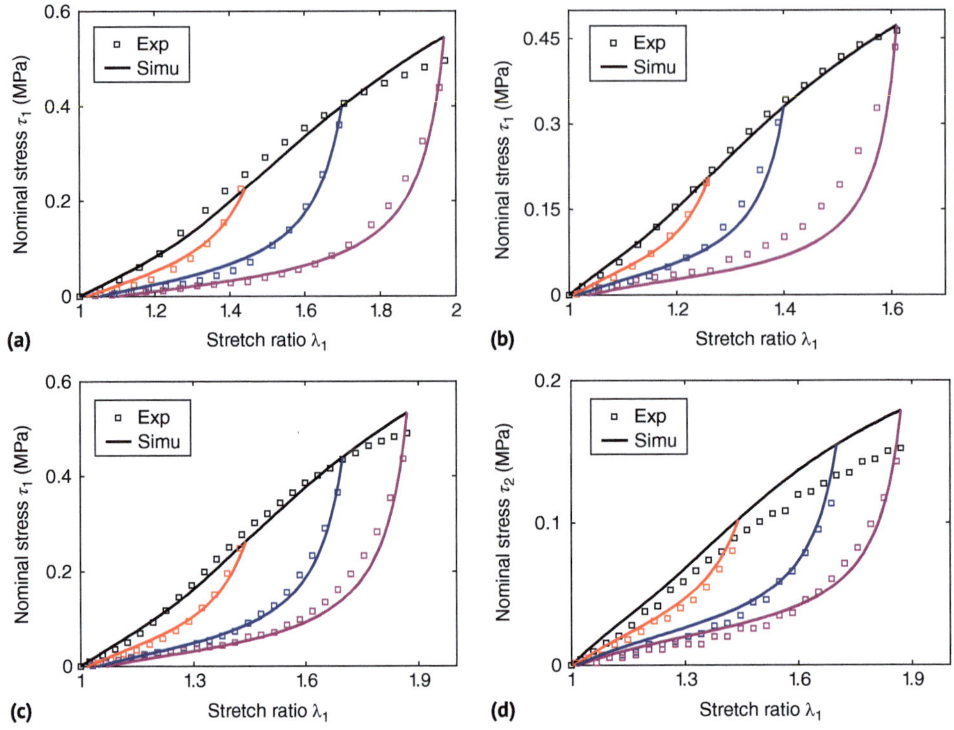

Figure 2.12 The performance of the non-affine damage model: (a) uniaxial tension, (b) equibiaxial tension, (c) pure shear (stress in $\mathbf{e}_1$ direction), and (d) pure shear (stress in $\mathbf{e}_2$ direction) with $G_0 = 0.037$ MPa, $N_0 = 1.33$, and $a_3 = 0.00062$.

Table 2.1 Pre-stretching ratio of multiple network elastomers.

Sample	λ_{s1}	λ_{s2}	λ_{s3}	λ_{s4}
Single network	1	NA	NA	NA
Double network 1	1.51	1	NA	NA
Double network 2	1.68	1	NA	NA
Triple network 1	2.18	1.65	1	NA
Triple network 2	2.41	1.58	1	NA
Triple network 3	2.55	1.65	1	NA
Quadruple network	3.42	2.27	1.39	1

Table 2.2 Model parameters for multiple network elastomers.

Parameter	Values	Physical significance
μ_f(MPa)	0.26	Shear modulus of the filler network
$\overline{E}_b$	5700	Normalized bond energy
λ_{cri}	1.18	Critical rupture ratio
μ_m(MPa)	0.1	Shear modulus of the matrix
N_m	100	Number of chain segments of the matrix

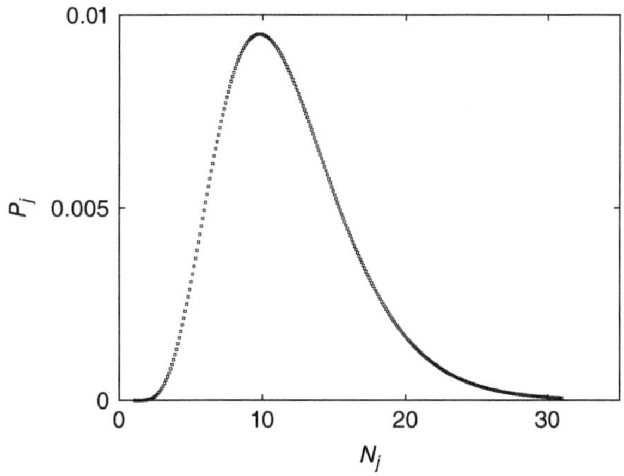

Figure 2.13 The distribution of chain lengths of the filler network.

2.5 Conclusion

Hyperelasticity is the most simple mechanical behavior of soft materials. The stress only depends on the deformation state. Thus, once the free energy function is defined, the explicit stress–strain relation is built. We have presented several representative hyperelastic models and demonstrated that it is necessary to incorporate the entanglement effect to fully capture the multi-axial stress response. The rate-dependence viscoelastic behaviors can be described by the rheological models. We show four viscoelastic models in finite deformation conditions through choosing different internal variables or configurations. Though these models are almost equivalent in the small strain condition, they exhibit different responses in large deformation. Thus, it is necessary to fully evaluate the performances of these models. The chapter further presented the classic damage models and two recently developed damage models for tough elastomers and gels. Understanding these complex nonlinear behaviors of soft materials can provide the basis for further development of the multi-field coupled models for stimulus-responsive soft materials.

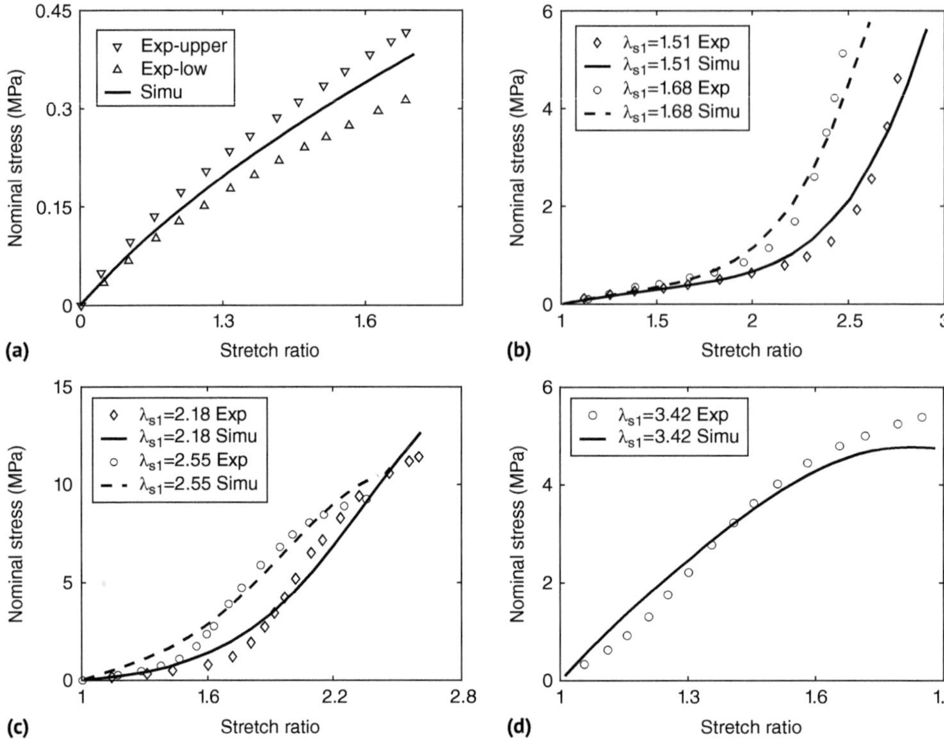

Figure 2.14 Comparing the measured and predicted stress response of (a) single network elastomers, (b) double network elastomers, (c) triple network elastomers, and (d) quadruple network elastomers. The experimental data is extracted from Millereau [59].

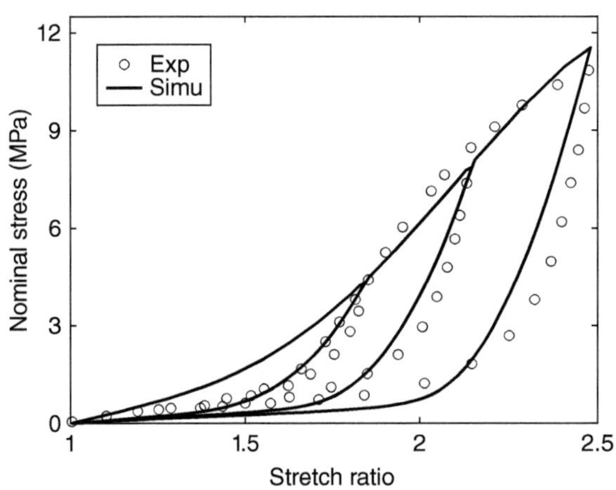

Figure 2.15 Comparing the measured and simulated stress response of triple network elastomers with $\lambda_{s1} = 2.41$ in cyclic tests. The experimental data is extracted from Millereau [59].

References

1 Kunquan Wang, Lu Dai, Ji Lin, Ping Rao, and Rui Xiao. Mechanical behaviors of copolymer elastomers: experimental characterization and constitutive modeling. *International Journal of Applied Mechanics*, 16(6): 2441002, 2024.

2 Melvin Mooney. A theory of large elastic deformation. *Journal of Applied Physics*, 11(9): 582–592, 1940.

3 Alan N Gent. A new constitutive relation for rubber. *Rubber Chemistry and Technology*, 69(1): 59–61, 1996.

4 Raymond William Ogden. Large deformation isotropic elasticity–on the correlation of theory and experiment for incompressible rubberlike solids. *Proceedings of the Royal Society of London. A. Mathematical and Physical Sciences*, 326(1567): 565–584, 1972.

5 M Kaliske and G Heinrich. An extended tube-model for rubber elasticity: statistical–mechanical theory and finite element implementation. *Rubber Chemistry and Technology*, 72(3): 602–632, 1999.

6 Mohd Shariff. Strain energy function for filled and unfilled rubberlike material. *Rubber Chemistry and Technology*, 73(1): 1–18, 2000.

7 Mario M Attard and Giles W Hunt. Hyperelastic constitutive modeling under finite strain. *International Journal of Solids and Structures*, 41(18–19): 5327–5350, 2004.

8 Xiang Yuhai, Zhong Danming, Wang Peng, Mao Guoyong, Yu Honghui, and Qu Shaoxing. A general constitutive model of soft elastomers. *Journal of the Mechanics and Physics of Solids*, 117: 110–122, 2018.

9 Vu Ngoc Khiêm and Mikhail Itskov. Analytical network-averaging of the tube model: rubber elasticity. *Journal of the Mechanics and Physics of Solids*, 95: 254–269, 2016.

10 Shun Meng, Haroon Imtiaz, and Bin Liu. A simple interpolation-based approach towards the development of an accurate phenomenological constitutive relation for isotropic hyperelastic materials. *Extreme Mechanics Letters*, 49: 101485, 2021.

11 Shiming Shen, Danming Zhong, Shaoxing Qu, and Rui Xiao. A hyperelastic-damage model based on the strain invariants. *Extreme Mechanics Letters*, 52: 101641, 2022.

12 Ming C Wang and Eugene Guth. Statistical theory of networks of non-Gaussian flexible chains. *The Journal of Chemical Physics*, 20(6): 1144–1157, 1952.

13 Ellen M Arruda and Mary C Boyce. A three-dimensional constitutive model for the large stretch behavior of rubber elastic materials. *Journal of the Mechanics and Physics of Solids*, 41(2): 389–412, 1993.

14 L Treloar and G Riding. A non-Gaussian theory for rubber in biaxial strain. I. Mechanical properties. *Proceedings of the Royal Society of London. Series A. Mathematical and Physical Sciences*, 369(1738): 261–280, 1979.

15 Jacob D Davidson and Nakhiah C Goulbourne. A nonaffine network model for elastomers undergoing finite deformations. *Journal of the Mechanics and Physics of Solids*, 61(8): 1784–1797, 2013.

16 Yuhai Xiang, Danming Zhong, Peng Wang, Tenghao Yin, Haofei Zhou, Honghui Yu, Chinmay Baliga, Shaoxing Qu, and Wei Yang. A physically based visco-hyperelastic constitutive model for soft materials. *Journal of the Mechanics and Physics of Solids*, 128: 208–218, 2019.

17 Hüsnü Dal, Osman Gültekin, and Kemal Açıkgöz. An extended eight-chain model for hyperelastic and finite viscoelastic response of rubberlike materials: Theory, experiments and numerical aspects. *Journal of the Mechanics and Physics of Solids*, 145: 104159, 2020.

18 Albert Edward Green and Ronald S Rivlin. The mechanics of non-linear materials with memory. *Archive for Rational Mechanics and Analysis*, 1(1): 1–21, 1957.

19 Albert E Green and Ronald S Rivlin. The mechanics of non-linear materials with memory. *Archive for Rational Mechanics and Analysis*, 4(1): 387–404, 1959.

20 Albert E Green. The equilibrium of rods. *Archive for Rational Mechanics and Analysis*, 3: 417–421, 1959.

21 Richard M Christensen and K H Lo. Solutions for effective shear properties in three phase sphere and cylinder models. *Journal of the Mechanics and Physics of Solids*, 27(4): 315–330, 1979.

22 Patrick Le Tallec, Christophe Rahier, and Ahmed Kaiss. Three-dimensional incompressible viscoelasticity in large strains: formulation and numerical approximation. *Computer Methods in Applied Mechanics and Engineering*, 109(3-4): 233–258, 1993.

23 Michael Kaliske and Heinrich Rothert. Formulation and implementation of three-dimensional viscoelasticity at small and finite strains. *Computational Mechanics*, 19(3): 228–239, 1997.

24 W L Vandoolaeghe and E M Terentjev. Constrained rouse model of rubber viscoelasticity. *The Journal of Chemical Physics*, 123(3): 034902, 2005.

25 W L Vandoolaeghe and E M Terentjev. A rouse-tube model of dynamic rubber viscoelasticity. *Journal of Physics A: Mathematical and Theoretical*, 40(49): 14725, 2007.

26 Ji Lin, Md Tariful Islam Mredha, Rumesh Rangana Manimel Wadu, Chuanqian Shi, Rui Xiao, Insu Jeon, and Jin Qian. Time-dependent constitutive behaviors of a dynamically crosslinked glycerogel governed by bond kinetics and chain diffusion. *Journal of the Mechanics and Physics of Solids*, 194: 105951, 2025.

27 Leonard Mullins. Effect of stretching on the properties of rubber. *Rubber Chemistry and Technology*, 21(2): 281–300, 1948.

28 Julie Diani, Bruno Fayolle, and Pierre Gilormini. A review on the mullins effect. *European Polymer Journal*, 45(3): 601–612, 2009.

29 Lin Zhan, Shaoxing Qu, and Rui Xiao. A review on the mullins effect in tough elastomers and gels. *Acta Mechanica Solida Sinica*, 37(2): 181–214, 2024.

30 Grégory Chagnon, Erwan Verron, Laurent Gornet, Gilles Marckmann, and Pierre Charrier. On the relevance of continuum damage mechanics as applied to the mullins effect in elastomers. *Journal of the Mechanics and Physics of Solids*, 52(7): 1627–1650, 2004.

31 Gilles Marckmann, Erwan Verron, Laurent Gornet, Gregory Chagnon, Pierre Charrier, and Patrice Fort. A theory of network alteration for the mullins effect. *Journal of the Mechanics and Physics of Solids*, 50(9): 2011–2028, 2002.

32 H J Qi and M C Boyce. Constitutive model for stretch-induced softening of the stress–stretch behavior of elastomeric materials. *Journal of the Mechanics and Physics of Solids*, 52(10): 2187–2205, 2004.

33 Xuanhe Zhao. A theory for large deformation and damage of interpenetrating polymer networks. *Journal of the Mechanics and Physics of Solids*, 60(2): 319–332, 2012.

34 Rui Xiao, Thanh-Tam Mai, Kenji Urayama, Jian Ping Gong, and Shaoxing Qu. Micromechanical modeling of the multi-axial deformation behavior in double network hydrogels. *International Journal of Plasticity*, 137: 102901, 2021b.

35 Leslie R G Treloar. Stress-strain data for vulcanized rubber under various types of deformation. *Rubber Chemistry and Technology*, 17(4): 813–825, 1944.

36 L R G Treloar. The elasticity of a network of long-chain molecules—II. *Transactions of the Faraday Society*, 39: 241–246, 1943.

37 Ronald S Rivlin. Large elastic deformations of isotropic materials IV. further developments of the general theory. *Philosophical Transactions of the Royal Society of London. Series A, Mathematical and Physical Sciences*, 241(835): 379–397, 1948.

38 Hüsnü Dal, Kemal Açıkgöz, and Yashar Badienia. On the performance of isotropic hyperelastic constitutive models for rubber-like materials: a state of the art review. *Applied Mechanics Reviews*, 73(2): 020802, 2021.

39 A G James, A Green, and G M Simpson. Strain energy functions of rubber. I. Characterization of gum vulcanizates. *Journal of Applied Polymer Science*, 19(7): 2033–2058, 1975.

40 P J Flory. Thermodynamic relations for high elastic materials. *Transactions of the Faraday Society*, 57: 829–838, 1961.

41 Rui Xiao, Ning Han, Danming Zhong, and Shaoxing Qu. Modeling the mechanical behaviors of multiple network elastomers. *Mechanics of Materials*, 161: 103992, 2021a.

42 Martin Kroeger. Simple, admissible, and accurate approximants of the inverse langevin and brillouin functions, relevant for strong polymer deformations and flows. *Journal of Non-Newtonian Fluid Mechanics*, 223: 77–87, 2015.

43 Masao Doi, Samuel Frederick Edwards, and Samuel Frederick Edwards. *The theory of polymer dynamics*, volume 73. Oxford University Press, 1988.

44 Víctor Jesús Amores, Khanh Nguyen, and Francisco Javier Montáns. On the network orientational affinity assumption in polymers and the micro–macro connection through the chain stretch. *Journal of the Mechanics and Physics of Solids*, 148: 104279, 2021.

45 Lin Zhan, Siyu Wang, Shaoxing Qu, Paul Steinmann, and Rui Xiao. A new micro–macro transition for hyperelastic materials. *Journal of the Mechanics and Physics of Solids*, 171: 105156, 2023.

46 Lingrui Zhu, Lin Zhan, and Rui Xiao. A comparative study of the entanglement models toward simulating hyperelastic behaviors. *Journal of Applied Mechanics*, 91(2): 021007, 2024.

47 Stefanie Reese and Sanjay Govindjee. A theory of finite viscoelasticity and numerical aspects. *International Journal of Solids and Structures*, 35(26-27): 3455–3482, 1998.

48 Christian Linder, Mykola Tkachuk, and Christian Miehe. A micromechanically motivated diffusion-based transient network model and its incorporation into finite rubber viscoelasticity. *Journal of the Mechanics and Physics of Solids*, 59(10): 2134–2156, 2011.

49 Franck J Vernerey, Rong Long, and Roberto Brighenti. A statistically-based continuum theory for polymers with transient networks. *Journal of the Mechanics and Physics of Solids*, 107: 1–20, 2017.

50 Juan Carlos Simo. On a fully three-dimensional finite-strain viscoelastic damage model: formulation and computational aspects. *Computer Methods in Applied Mechanics and Engineering*, 60(2): 153–173, 1987.

51 Richard M Christensen. *Theory of viscoelasticity*. Courier Corporation, 2013.

52 Danming Zhong, Yuhai Xiang, Zhicheng Wang, Zhe Chen, Junjie Liu, Zi Liang Wu, Rui Xiao, Shaoxing Qu, and Wei Yang. A visco-hyperelastic model for hydrogels with tunable water content. *Journal of the Mechanics and Physics of Solids*, 173: 105206, 2023.

53 Sanjay Govindjee and Stefanie Reese. A presentation and comparison of two large deformation viscoelasticity models. *Journal of Engineering Materials and Technology*, 119: 251–255, 1997.

54 Christian Miehe. Discontinuous and continuous damage evolution in ogden-type large-strain elastic materials. *European Journal of Mechanics. A. Solids*, 14(5): 697–720, 1995.

55 Zdeněk P Bažant and Byung H Oh. Efficient numerical integration on the surface of a sphere. *ZAMM-Journal of Applied Mathematics and Mechanics*, 66(1): 37–49, 1986.

56 Yunwei Mao, Brandon Talamini, and Lallit Anand. Rupture of polymers by chain scission. *Extreme Mechanics Letters*, 13: 17–24, 2017.

57 Thanh-Tam Mai, Takahiro Matsuda, Tasuku Nakajima, Jian Ping Gong, and Kenji Urayama. Distinctive characteristics of internal fracture in tough double network hydrogels revealed by various modes of stretching. *Macromolecules*, 51(14): 5245–5257, 2018.

58 Sanjay Govindjee and Juan Simo. A micro-mechanically based continuum damage model for carbon black-filled rubbers incorporating mullins' effect. *Journal of the Mechanics and Physics of Solids*, 39(1): 87–112, 1991.

59 Pierre Michel Millereau. *Large Strain and Fracture of Multiple Network Elastomers*. PhD thesis, Université Pierre et Marie Curie-Paris VI, 2017.

3

A Thermomechanical Coupled Model for Amorphous Shape-memory Polymers

Shape-memory polymers (SMPs) are an important type of soft active materials, which have received considerable research interest in the past decades. In this chapter, we present an effective temperature model to describe the thermo-mechanical behaviors and shape-memory behaviors of amorphous polymers. It is shown that the classic finite deformation viscoelastic model is a special case of the effective temperature theory. Though the simple viscoelastic model is able to capture many features of the shape-memory effect, the effective temperature theory is proven to serve as a more unified theoretical framework. This framework can capture the stress response of amorphous polymers across the glass transition region. It can also predict the influences of aging on the shape recovery performance. The model is also applied to investigate the stress response in the constrained recovery tests for both hot-programmed and cold-programmed SMPs. The model is further extended to simulate the solvent-driven shape-memory behaviors in amorphous polymeric materials. The chapter also discusses the remaining challenges for modeling shape-memory effects in polymers.

3.1 Introduction

Shape-memory polymers (SMPs) can recover from a temporary shape to a permanent shape upon an external stimulus [1–3]. The permanent shape of SMPs is typically determined by cross-linked networks or heavily entangled networks. A temporary shape is mainly preprogrammed through mechanical deformation at a relatively high temperature and fixed through cooling to a low temperature. This is due to a tremendous change in chain mobility induced by phase transition, either a glass transition for amorphous polymers or a melting transition for semicrystalline polymers [4]. In this sense, most polymer systems can exhibit a certain level of shape-memory performance, even for traditional engineering polymers, such as natural rubber [5] and thermoplastics [6]. Nevertheless, the shape-memory performance, such as the shape-fixity ratio, shape-recovery ratio, still strongly depends on the material types [7].

The shape-memory behaviors can be characterized through either measuring the strain recovery in the free recovery state or stress response in the fixed-strain recovery process. In addition to the chemical composition, programming and recovery conditions also significantly affect the responses. For example, both amorphous SMPs and semicrystalline SMPs

Multi-field Modeling of Soft Active Materials: Properties and Design, First Edition. Rui Xiao.

exhibit the so-called temperature memory effect, namely, a higher programmed temperature resulting in a higher recovery temperature [8, 9]. Ortega et al. [10] showed that long-term storage can shift the shape recovery region to a higher temperature and also induce a larger initial recovery rate compared with SMPs without storage. This is caused by the aging effect, which induces a more sluggish polymer structure during storage. Shape recovery can be achieved in a continuous manner at a constant heating rate or in a discrete manner at multiple temperatures [11]. For semicrystalline polymers, the reversible two-way shape-memory effect can also be achieved through a specific design [12].

It is highly desirable to develop constitutive models to describe the thermomechanical properties and shape-memory behaviors in polymers. The models can assist the design of novel SMPs and promote the practical applications of these active polymeric materials. Due to different mechanisms for shape-memory effects in amorphous polymers and semicrystalline polymers, the theoretical framework also varies [13]. For amorphous polymers, the glass transition mechanism needs to be incorporated into the model [14]. Yu and Qi [15] developed a one-dimensional viscoelastic model through an extension on the generalized Maxwell model by assuming temperature-dependent viscosity. The model is able to describe the temperature-memory effect observed in amorphous polymers. Diani et al. [16] constructed a finite deformation viscoelastic model for SMPs. The model is able to simulate the torsion shape recovery of epoxy under various heating rates. The viscoelastic model can be further extended to the viscoplastic condition, as demonstrated by Nguyen et al. [14] and Srivastava et al. [17]. In order to capture the aging effect, additional internal variables, such as fictive temperature, free volume need to be included in the constitutive model [18, 19]. For semicrystalline polymers, the melting transition needs to be incorporated into the model [20, 21]. These models need to specify the constitutive relations for different phases, and also the transition between two phases as temperature changes. Though the melting transition should be the main mechanism for shape-memory effects in semicrystalline polymers, these models can also well predict the shape-memory behaviors in amorphous polymeric materials [20]. The phase transition models can also be extended to model the two-way shape-memory effect [22–25].

In addition to the thermally activated process, solvent-induced shape-memory effects have been reported for specific material systems. The programmed shape can recover in water, organic solvent or exposure to moisture in a thermal condition [4]. Several constitutive models have been developed for these material systems [26–28]. To capture the solvent-driven shape-memory behaviors, the models need to introduce the effects of solvent on the properties. In addition, the diffusion process needs to be included [26, 27]. Xiao and Nguyen [26] developed a chemo-mechanical coupled model that can capture the stress response of both dry and wet polymers and also the shape recovery of programmed SMP samples in water. Light-induced shape recovery has also been reported in several material systems [29]. Long et al. [30] developed constitutive models for light-activated SMPs through incorporation of the chemo-mechanical coupling effect, which reproduces the main features of the experimental findings.

As discussed above, many models have been developed for SMPs. It is difficult to cover all these progress in one single chapter. Thus, this chapter focuses on amorphous SMPs, which have been investigated extensively by our group [26, 28, 31–34]. The remaining part of the chapter is arranged as follows. In Section 3.2, we first introduce a thermodynamic framework for the effective temperature theory, which can be treated as a unified model for amorphous polymers across the glass transition region. The explicit form is then specified, followed by a viscoelastic model as a simplified version. Section 3.3 briefly discusses the procedures of determining model parameters for the effective temperature theory. Section 3.4 presents the simulation results from both the viscoelastic model and effective temperature theory. The model has also been demonstrated to capture the solvent-driven shape-memory effect through certain extensions. Section 3.5 discusses the current challenges for modeling shape-memory effect in polymers. The chapter ends with a conclusion.

3.2 Thermodynamics

3.2.1 Kinematics

Though the kinematics have been introduced in the previous chapter, we still briefly summarize the main definitions for the convenience of reading. The deformation gradient $\mathbf{F}$ is defined to map the point $\mathbf{X}$ in the reference undeformed configuration to the point $\mathbf{x}(\mathbf{X})$ in the current deformed configuration as $\mathbf{F} = \partial \mathbf{x}(\mathbf{X}) / \partial \mathbf{X}$.

The deformation gradient is decomposed into multiple pairs of elastic and viscous deformation gradients as

$$\mathbf{F} = \mathbf{F}_j^{\mathrm{e}} \mathbf{F}_j^{\mathrm{v}}, \quad j = 1, \ldots, N. \tag{3.1}$$

The deformation gradient is also split into volumetric and deviatoric parts as

$$\overline{\mathbf{F}} = J^{-1/3} \mathbf{F}, \quad \overline{\mathbf{F}}_j^{\mathrm{e}} = J_j^{\mathrm{e}\text{-}1/3} \mathbf{F}_j^{\mathrm{e}}, \tag{3.2}$$

where $J = \det(\mathbf{F})$ and $J_j^{\mathrm{e}} = \det\left(\mathbf{F}_j^{\mathrm{e}}\right)$.

The right and left Cauchy deformation tensors are then defined as

$$\begin{aligned} \mathbf{C} &= \mathbf{F}^{\mathrm{T}} \mathbf{F}, \quad \mathbf{C}_j^{\mathrm{e}} = \mathbf{F}_j^{\mathrm{eT}} \mathbf{F}_j^{\mathrm{e}}, \\ \mathbf{b} &= \mathbf{F} \mathbf{F}^{\mathrm{T}}, \quad \mathbf{b}_j^{\mathrm{e}} = \mathbf{F}_j^{\mathrm{e}} \mathbf{F}_j^{\mathrm{eT}}. \end{aligned} \tag{3.3}$$

The corresponding deviatoric parts of Cauchy deformation tensor are defined as

$$\begin{aligned} \overline{\mathbf{C}} &= J^{-2/3} \mathbf{C}, \quad \overline{\mathbf{C}}_j^{\mathrm{e}} = J_j^{\mathrm{e}\text{-}2/3} \mathbf{C}_j^{\mathrm{e}}, \\ \overline{\mathbf{b}} &= J^{-2/3} \mathbf{b}, \quad \overline{\mathbf{b}}_j^{\mathrm{e}} = J_j^{\mathrm{e}\text{-}2/3} \mathbf{b}_j^{\mathrm{e}}. \end{aligned} \tag{3.4}$$

The velocity gradients are defined as

$$\mathbf{L} = \dot{\mathbf{F}} \mathbf{F}^{-1}, \quad \mathbf{L}_j^{\mathrm{v}} = \dot{\mathbf{F}}_j^{\mathrm{v}} \mathbf{F}_j^{\mathrm{v}\text{-}1}. \tag{3.5}$$

The symmetric parts of the elastic and viscous velocity gradients are further defined as $\mathbf{D} = \left(\mathbf{L} + \mathbf{L}^{\mathrm{T}}\right)/2$ and $\mathbf{D}_j^{\mathrm{v}} = \left(\mathbf{L}_j^{\mathrm{v}} + \mathbf{L}_j^{\mathrm{vT}}\right)/2$.

3.2.2 Thermodynamic Framework

This part extends the thermodynamic framework in Chapters 1 and 2 to the nonequilibrium condition by introducing two subsystems, a fast kinetic subsystem and a slow configurational subsystem [35–37]. The slow configurational subsystem consists of multiple components due to the broad distribution of relaxation time.

The first law of thermodynamics is represented as

$$\dot{\mathrm{e}} = \mathbf{S} : \frac{1}{2}\dot{\mathbf{C}} - \nabla_{\mathbf{X}} \cdot \mathbf{Q}, \tag{3.6}$$

where e is the internal energy, **S** is the second Piola–Kirchhoff stress tensor, and **Q** is the heat flux flowing into the material systems.

The second law of thermodynamics requires

$$\dot{\eta} + \nabla_{\mathbf{X}} \cdot \mathbf{H} \geq 0, \tag{3.7}$$

where η is the entropy of the system and **H** is the entropy flux.

The thermodynamic variables are split into a component of the kinetic subsystem and components of the configurational subsystems, represented as

$$\begin{aligned} &\mathrm{e} = \mathrm{e}^k + \sum_i^P \mathrm{e}_i^c, \quad \eta = \eta^{\mathrm{k}} + \sum_i^P \eta_i^c, \\ &\mathbf{Q} = \mathbf{Q}^k + \sum_i^P \mathbf{Q}_i^c, \quad \mathbf{H} = \mathbf{H}^k + \sum_i^P \mathbf{H}_i^c. \end{aligned} \tag{3.8}$$

The entropy flux in different subsystems is then defined as

$$\mathbf{H}^{\mathrm{k}} = \frac{\mathbf{Q}^k}{T}, \quad \mathbf{H}_i^{\mathrm{c}} = \frac{\mathbf{Q}_i^c}{T_{e_i}}, \tag{3.9}$$

where T_{e_i} is the effective temperature of ith configurational subsystem.

The Helmholtz free energy density is represented by Nieuwenhuizen [35] as

$$\begin{aligned} &\Psi = \Psi^{\mathrm{k}} + \sum_i^P \Psi_i^{\mathrm{c}}, \\ &\Psi^{\mathrm{k}} = \mathrm{e}^k - T\eta^{\mathrm{k}}, \\ &\Psi_i^{\mathrm{c}} = \mathrm{e}_i^c - T_{e_i}\eta_i^c, \end{aligned} \tag{3.10}$$

where Ψ^{k} and Ψ_i^{c} are the Helmholtz free energy density contributed by the kinetic and configurational parts, respectively.

The first law can be rewritten by combining Eqs. (3.6) and (3.10) as

$$\dot{\Psi} + T\dot{\eta}^{\mathrm{k}} + \sum_i^P T_{e_i}\dot{\eta}_i^c + \eta^{\mathrm{k}}\dot{T} + \sum_i^P \eta_i^{\mathrm{c}}\dot{T}_{e_i} = \mathbf{S} : \frac{1}{2}\dot{\mathbf{C}} - \nabla_{\mathbf{X}} \cdot \mathbf{Q}^{\mathrm{k}} - \sum_i^P \nabla_{\mathbf{X}} \cdot \mathbf{Q}_i^{\mathrm{c}}. \tag{3.11}$$

The second law of thermodynamics can be further expanded in the following form through substituting Eqs. (3.8) and (3.9) into Eq. (3.7):

$$\dot{\eta}^{\mathrm{k}}+\sum_{i}^{P}\dot{\eta}_i^{\mathrm{c}}+\frac{\nabla_{\mathbf{X}}\cdot\mathbf{Q}^{\mathrm{k}}}{T}-\frac{1}{T^2}\mathbf{Q}^{\mathrm{k}}\cdot\nabla_{\mathbf{X}}T+\sum_{i}^{P}\frac{\nabla_{\mathbf{X}}\cdot\mathbf{Q}_i^{\mathrm{c}}}{T_{e_i}}-\sum_{i}^{P}\frac{1}{T_{e_i}^2}\mathbf{Q}_i^{\mathrm{c}}\cdot\nabla_{\mathbf{X}}T_{e_i}\geq 0. \tag{3.12}$$

Combining Eqs. (3.11) and (3.12) to eliminate the kinetic entropy η^{k} results in the following form for the entropy inequality:

$$\begin{aligned}&\sum_{i}^{P}(T-T_{e_i})\dot{\eta}_i^{\mathrm{c}}+\sum_{i}^{P}(T-T_{e_i})\frac{\nabla_{\mathbf{X}}\cdot\mathbf{Q}_i^{\mathrm{c}}}{T_{e_i}}-\frac{1}{T}\mathbf{Q}^{\mathrm{k}}\cdot\nabla_{\mathbf{X}}T-\sum_{i}^{P}\frac{T}{T_{e_i}^2}\mathbf{Q}_i^{\mathrm{c}}\cdot\nabla_{\mathbf{X}}T_{e_i}\\&+\mathbf{S}:\frac{1}{2}\dot{\mathbf{C}}-\eta^{\mathrm{k}}\dot{T}-\sum_{i}^{P}\eta_i^{\mathrm{c}}\dot{T}_{e_i}-\dot{\Psi}\geq 0.\end{aligned} \tag{3.13}$$

The Helmholtz free energy density is assumed to be dependent on the total and elastic deformation tensors $\mathbf{C}$ and $\mathbf{C}_j^{\mathrm{e}}$, temperature T, effective temperatures T_{e_i}, and the internal variables $\mathbf{A}_j$ representing the orientation tensor for strain hardening responses as

$$\Psi=\Psi\left(T,T_{e_i},\vec{\mathbf{C}},\vec{\mathbf{C}}_j^{\mathrm{e}},\vec{\mathbf{A}}_j\right). \tag{3.14}$$

The rate of the Helmholtz free energy density is expanded as

$$\dot{\Psi}=\frac{\partial\Psi}{\partial T}\dot{T}+\sum_{i}^{P}\frac{\partial\Psi}{\partial T_{e_i}}\dot{T}_{e_i}+2\frac{\partial\Psi}{\partial\mathbf{C}}:\frac{1}{2}\dot{\mathbf{C}}-\sum_{j}^{N}2\mathbf{C}_j^{\mathrm{e}}\frac{\partial\Psi}{\partial\mathbf{C}_j^{\mathrm{e}}}:\mathbf{D}_j^{\mathrm{v}}+\sum_{j}^{N}\frac{\partial\Psi}{\partial\mathbf{A}_j}:\dot{\mathbf{A}}_j. \tag{3.15}$$

Substituting Eq. (3.15) into Eq. (3.13) yields

$$\begin{aligned}&\sum_{i}^{P}(T-T_{e_i})\dot{\eta}_i^{\mathrm{c}}+\sum_{i}^{P}(T-T_{e_i})\frac{\nabla_{\mathbf{X}}\cdot\mathbf{Q}_i^{\mathrm{c}}}{T_{e_i}}-\frac{1}{T}\mathbf{Q}^{\mathrm{k}}\cdot\nabla_{\mathbf{X}}T-\sum_{i}^{P}\frac{T}{T_{e_i}^2}\mathbf{Q}_i^{\mathrm{c}}\cdot\nabla_{\mathbf{X}}T_{e_i}\\&-\sum_{j}^{N}\frac{\partial\Psi}{\partial\mathbf{A}_j}:\dot{\mathbf{A}}_j+\sum_{j}^{N}2\mathbf{C}_j^{\mathrm{e}}\frac{\partial\Psi}{\partial\mathbf{C}_j^{\mathrm{e}}}:\mathbf{D}_j^{\mathrm{v}}+\left(\mathbf{S}-2\frac{\partial\Psi}{\partial\mathbf{C}}\right):\frac{1}{2}\dot{\mathbf{C}}-\left(\eta^{\mathrm{k}}+\frac{\partial\Psi}{\partial T}\right)\dot{T}\\&-\sum_{i}^{P}\left(\eta_i^{\mathrm{c}}+\frac{\partial\Psi}{\partial T_{e_i}}\right)\dot{T}_{e_i}\geq 0.\end{aligned} \tag{3.16}$$

The equilibrium thermodynamics gives

$$\mathbf{S}=2\frac{\partial\Psi}{\partial\mathbf{C}},\quad \eta^{\mathrm{k}}=-\frac{\partial\Psi}{\partial T},\quad \eta_i^{\mathrm{c}}=-\frac{\partial\Psi}{\partial T_{e_i}}. \tag{3.17}$$

Thus, the second law (Eq. 3.16) can be further reduced to

$$\begin{aligned}&\sum_{i}^{P}(T-T_{e_i})\dot{\eta}_i^{\mathrm{c}}+\sum_{i}^{P}(T-T_{e_i})\frac{\nabla_{\mathbf{X}}\cdot\mathbf{Q}_i^{\mathrm{c}}}{T_{e_i}}-\frac{1}{T}\mathbf{Q}^{\mathrm{k}}\cdot\nabla_{\mathbf{X}}T-\sum_{i}^{P}\frac{T}{T_{e_i}^2}\mathbf{Q}_i^{\mathrm{c}}\cdot\nabla_{\mathbf{X}}T_{e_i}\\&-\sum_{j}^{N}\frac{\partial\Psi}{\partial\mathbf{A}_j}:\dot{\mathbf{A}}_j+\sum_{j}^{N}2\mathbf{C}_j^{\mathrm{e}}\frac{\partial\Psi}{\partial\mathbf{C}_j^{\mathrm{e}}}:\mathbf{D}_j^{\mathrm{v}}\geq 0.\end{aligned} \tag{3.18}$$

The following form is adopted for the evolution equation of the orientation tensor $\mathbf{A}_j$ [38, 39]:

$$\dot{\mathbf{A}}_j = \mathbf{A}_j\mathbf{D}_j^{\mathrm{v}} + \mathbf{D}_j^{\mathrm{v}}\mathbf{A}_j - \boldsymbol{\phi}(\mathbf{A}_j), \tag{3.19}$$

where $\boldsymbol{\phi}(\mathbf{A}_j)$ is a relaxation term for the orientation tensor.

Equation (3.19) is then substituted into Eq. (3.18), which gives

$$\begin{aligned}&\sum_i^P (T - T_{e_i})\dot{\eta}_i^{\mathrm{c}} + \sum_i^P (T - T_{e_i})\frac{\nabla_{\mathbf{X}} \cdot \mathbf{Q}_i^{\mathrm{c}}}{T_{e_i}} - \frac{1}{T}\mathbf{Q}^{\mathrm{k}} \cdot \nabla_{\mathbf{X}} T - \sum_i^P \frac{T}{T_{e_i}^2}\mathbf{Q}_i^{\mathrm{c}} \cdot \nabla_{\mathbf{X}} T_{e_i}\\ &+ \sum_j^N \frac{\partial \Psi}{\partial \mathbf{A}_j} : \boldsymbol{\phi}(\mathbf{A}_j) + \sum_j^N \left(2\mathbf{C}_j^{\mathrm{e}}\frac{\partial \Psi}{\partial \mathbf{C}_j^{\mathrm{e}}} - 2\mathbf{A}_j\frac{\partial \Psi}{\partial \mathbf{A}_j}\right) : \mathbf{D}_j^{\mathrm{v}} \geq 0.\end{aligned} \tag{3.20}$$

The last two terms on the right side of Eq. (3.20) are grouped together as the dissipative energy

$$\mathcal{W}^{\mathrm{int}} = \sum_j^N \frac{\partial \Psi}{\partial \mathbf{A}_j} : \boldsymbol{\phi}(\mathbf{A}_j) + \sum_j^N \left(2\mathbf{C}_j^{\mathrm{e}}\frac{\partial \Psi}{\partial \mathbf{C}_j^{\mathrm{e}}} - 2\mathbf{A}_j\frac{\partial \Psi}{\partial \mathbf{A}_j}\right) : \mathbf{D}_j^{\mathrm{v}}. \tag{3.21}$$

Combining Eqs. (3.11), (3.15), (3.17), (3.19), and (3.21) gives a new expression for the first law:

$$T\dot{\eta}^{\mathrm{k}} + \sum_i^P T_{e_i}\dot{\eta}_i^{\mathrm{c}} = -\nabla_{\mathbf{X}} \cdot \mathbf{Q}^{\mathrm{k}} - \sum_i^P \nabla_{\mathbf{X}} \cdot \mathbf{Q}_i^{\mathrm{c}} + \mathcal{W}^{\mathrm{int}}. \tag{3.22}$$

Several assumptions are made to obtain the corresponding first law of kinetic and configurational subsystems. First, in addition to the heat exchange within each subsystem, the kinetic subsystem and each configurational subsystem have a heat exchange denoted as Q_i^{kc}, while the related heat exchange between different configurational subsystems is ignored [40]. Second, the dissipative energy is distributed into all the subsystems represented as $\mathcal{W}^{\mathrm{int}} = \mathcal{W}^{\mathrm{int,k}} + \sum_i^P \mathcal{W}_i^{\mathrm{int,c}}$, where $\mathcal{W}^{\mathrm{int,k}} = \sum_j^N \frac{\partial \Psi^{\mathrm{k}}}{\partial \mathbf{A}_j} : \boldsymbol{\phi}(\mathbf{A}_j) + \sum_j^N \left(2\mathbf{C}_j^{\mathrm{e}}\frac{\partial \Psi^{\mathrm{k}}}{\partial \mathbf{C}_j^{\mathrm{e}}} - 2\mathbf{A}_j\frac{\partial \Psi^{\mathrm{k}}}{\partial \mathbf{A}_j}\right) : \mathbf{D}_j^{\mathrm{v}}$ and $\mathcal{W}_i^{\mathrm{int,c}} = \sum_j^N \frac{\partial \Psi_i^{\mathrm{c}}}{\partial \mathbf{A}_j} : \boldsymbol{\phi}(\mathbf{A}_j) + \sum_j^N (2\mathbf{C}_j^{\mathrm{e}}\frac{\partial \Psi_i^{\mathrm{c}}}{\partial \mathbf{C}_j^{\mathrm{e}}} - 2\mathbf{A}_j\frac{\partial \Psi_i^{\mathrm{c}}}{\partial \mathbf{A}_j}) : \mathbf{D}_j^{\mathrm{v}}$ [37, 40].

The first law in each subsystem can be represented as

$$\begin{aligned}T\dot{\eta}^{\mathrm{k}} &= -\nabla_{\mathbf{X}} \cdot \mathbf{Q}^{\mathrm{k}} - \sum_i^P Q_i^{kc} + \mathcal{W}^{\mathrm{int},k},\\ T_{e_i}\dot{\eta}_i^{\mathrm{c}} &= -\nabla_{\mathbf{X}} \cdot \mathbf{Q}_i^{\mathrm{c}} + Q_i^{kc} + \mathcal{W}_i^{\mathrm{int},c}, \quad i = 1, \ldots, P.\end{aligned} \tag{3.23}$$

The second equation in Eq. (3.23) is then used to eliminate the η_i^{c} term in Eq. (3.20), which gives

$$-\frac{1}{T}\mathbf{Q}^{\mathrm{k}} \cdot \nabla_{\mathbf{X}} T - \sum_i^P \frac{T}{T_{e_i}^2}\mathbf{Q}_i^{\mathrm{c}} \cdot \nabla_{\mathbf{X}} T_{e_i} + \sum_i^P (T - T_{e_i})\frac{Q_i^{kc}}{T_{e_i}} + \mathcal{W}^{\mathrm{int},k} + \sum_i^p \frac{T}{T_{e_i}}\mathcal{W}_i^{\mathrm{int},c} \geq 0. \tag{3.24}$$

Here, each term in Eq. (3.24) is required to be nonnegative as

$$\begin{aligned}
&-\frac{1}{T}\mathbf{Q}^{\mathrm{k}}\cdot\nabla_{\mathbf{X}}T\geq 0,\\
&-\frac{T}{T_{e_i}^2}\mathbf{Q}_i^{\mathrm{c}}\cdot\nabla_{\mathbf{X}}T_{e_i}\geq 0,\\
&\left(T-T_{e_i}\right)\frac{Q_i^{kc}}{T_{e_i}}\geq 0,\\
&\mathcal{W}^{\mathrm{int},k}\geq 0,\\
&\frac{T}{T_{e_i}}\mathcal{W}_i^{\mathrm{int},c}\geq 0.
\end{aligned} \tag{3.25}$$

The first two components in Eq. (3.25) represent the entropy inequality caused by heat flux, which can be satisfied by using a linear Fourier-type conduction equation as

$$\mathbf{q}^{\mathrm{k}}=-k^{\mathrm{k}}\nabla_{\mathbf{x}}T,\quad \mathbf{q}_i^{\mathrm{c}}=-k_i^{\mathrm{c}}\nabla_{\mathbf{x}}T_{e_i}, \tag{3.26}$$

where $\mathbf{q}^{\mathrm{k}}=1/J\mathbf{FQ}$ and $\mathbf{q}_i^{\mathrm{c}}=1/J\mathbf{FQ}_i^{\mathrm{c}}$ are the heat fluxes contributed by the kinetic and configurational subsystem in the current configuration, and the nonnegative variables of k^{k} and k_i^{c} are the corresponding thermal conductivity of each subsystem [41].

The third equation in Eq. (3.25) indicates that a heat flux occurs when the configurational subsystem is out of equilibrium ($T\neq T_{e_i}$). This heat flux drives the effective temperature toward temperature, which is the underlying physical mechanism for physical aging. This inequality can be satisfied by adopting the following equation for Q_i^{kc},

$$Q_i^{kc}=K_i\left(T-T_{e_i}\right),\quad i=1,\ \ldots,\ P, \tag{3.27}$$

where $K_i\geq 0$ is a parameter similar to the heat transfer coefficient.

The last two equations of Eq. (3.25) require that the dissipative energy flowing into either the kinetic or configurational subsystem cannot be negative, which automatically requires $\mathcal{W}^{\mathrm{int}}\geq 0$. This can be satisfied through

$$\begin{aligned}
&\left(2\mathbf{C}_j^{\mathrm{e}}\frac{\partial\Psi}{\partial\mathbf{C}_j^{\mathrm{e}}}-2\mathbf{A}_j\frac{\partial\Psi}{\partial\mathbf{A}_j}\right):\mathbf{D}_j^{\mathrm{v}}\geq 0,\\
&\frac{\partial\Psi}{\partial\mathbf{A}_j}:\boldsymbol{\phi}\left(\mathbf{A}_j\right)\geq 0.
\end{aligned} \tag{3.28}$$

The following relation is then adopted for $\mathbf{D}_j^{\mathrm{v}}$ [39]:

$$\mathbf{D}_j^{\mathrm{v}}=\frac{\mathbf{M}_j^{\mathrm{eff}}}{2\upsilon_j},\quad \mathbf{M}_j^{\mathrm{eff}}=\underbrace{2\mathbf{C}_j^{\mathrm{e}}\frac{\partial\Psi}{\partial\mathbf{C}_j^{\mathrm{e}}}}_{\mathbf{M}_j^{\mathrm{neq}}}-\underbrace{2\mathbf{A}_j\frac{\partial\Psi}{\partial\mathbf{A}_j}}_{\mathbf{M}_j^{\mathrm{back}}},\quad j=1\ldots N, \tag{3.29}$$

where $\mathbf{M}_j^{\mathrm{neq}}$ is the Mandel stress, $\mathbf{M}_j^{\mathrm{back}}$ is the back stress accounting for the strain-hardening response and $\mathbf{M}_j^{\mathrm{eff}}$ is the effective Mandel stress for the viscous deformation.

The following equation is used to meet the requirement of the nonnegative dissipative energy for chain orientation [39]:

$$\phi\left(\mathbf{A}_j\right) = \frac{2}{\upsilon_j^{\text{ori}}}\mathbf{A}_j\frac{\partial\Psi}{\partial\mathbf{A}_j}\mathbf{A}_j, \tag{3.30}$$

where $\upsilon_j^{\text{ori}} > 0$. The dissipative energy contributed by the relaxation of chain orientation can then be written as

$$\frac{\partial\Psi}{\partial\mathbf{A}_j} : \phi\left(\mathbf{A}_j\right) = \frac{1}{2\upsilon_j^{\text{ori}}}\mathbf{M}_j^{\text{back}} : \mathbf{M}_j^{\text{back}} \geq 0. \tag{3.31}$$

3.2.3 Constitutive Relationships

The Helmholtz free energy density is represented in the following form:

$$\begin{aligned}
\Psi^{\text{k}} = {} & (1-a)\sum_j^N\left\{\frac{\mu_j^{\text{neq}}}{2}\left(\text{tr}\left(\overline{\mathbf{C}}_j^{\text{e}}\right)-3\right)\right\} \\
& + (1-a)\sum_j^N\left\{\mu_j^{\text{ori}}\lambda_{\text{L}}^2\left[\frac{\overline{\lambda}_j^{\text{eff}}}{\lambda_{\text{L}}}x+\ln\left(\frac{x}{\sinh x}\right)\right]-\frac{1}{\lambda_{\text{L}}}y-\ln\left(\frac{y}{\sinh y}\right)\right\} \\
& + \frac{\kappa}{4}\left(J^2-2\log J-1\right)+c_{\text{g}}\left(T-T_0\right)-c_{\text{g}}T\ln\frac{T}{T_0}, \\
\Psi_i^{\text{c}} = {} & a\phi_i\sum_j^N\left\{\frac{\mu_j^{\text{neq}}}{2}\left(\text{tr}\left(\overline{\mathbf{C}}_j^{\text{e}}\right)-3\right)\right\} \\
& + a\phi_i\sum_j^N\left\{\mu_j^{\text{ori}}\lambda_{\text{L}}^2\left[\frac{\overline{\lambda}_j^{\text{eff}}}{\lambda_{\text{L}}}x+\ln\left(\frac{x}{\sinh x}\right)-\frac{1}{\lambda_{\text{L}}}y-\ln\left(\frac{y}{\sinh y}\right)\right]\right\} \\
& + \phi_i\frac{T_{e_i}}{T_0}\frac{\mu^{\text{eq}}}{2}\left(\text{tr}\left(\overline{\mathbf{C}}\right)-3\right)+\phi_i\Delta c\left(T_{e_i}-T_0\right)-\phi_i\Delta cT_{e_i}\ln\frac{T_{e_i}}{T_2},
\end{aligned} \tag{3.32}$$

where μ^{eq} is the equilibrium shear modulus, μ_j^{neq} is the nonequilibrium shear modulus of jth process, κ is the bulk modulus, μ_j^{ori} is the stiffness of orientation of polymer chain, λ_{L} is the stretch limit of the polymer chain, c_{g} is the heat capacity of the kinetic subsystem, Δc is the total heat capacity of the configurational subsystems, T_2 is the Kauzmann temperature, T_0 is the reference temperature, the parameter a with a restriction of $0 \leq a \leq 1$ is a partition parameter representing the fraction of dissipative energy flowing into the configurational subsystems, and ϕ_i represents the distribution of the configurational subsystems with $0 < \phi_i \leq 1$ and $\sum_i^P \phi_i = 1$. Here, the eight-chain model [42] is employed for the orientation of the polymer network for the strain-hardening effect with $\overline{\lambda}_j^{\text{eff}} = \sqrt{\text{tr}\overline{\mathbf{A}}_j/3}$, $x = \mathcal{L}^{-1}\left(\overline{\lambda}_j^{\text{eff}}/\lambda_{\text{L}}\right)$ and $y = \mathcal{L}^{-1}\left(1/\lambda_{\text{L}}\right)$, where $\mathcal{L}^{-1}(\cdot)$ is the inverse Langevin function.

Based on Eqs. (3.32) and (3.17), the second Piola–Kirchhoff stress tensor, the kinetic entropy and the configurational entropy can be calculated as

$$\begin{aligned} \mathbf{S} &= \sum_i^P \phi_i \frac{T_{e_i}}{T_0} \mu^{\mathrm{eq}} J^{-2/3} \left(\mathbf{I} - \frac{1}{3} \mathrm{tr}(\mathbf{C}) \mathbf{C}^{-1} \right) \\ &\quad + \sum_j^N \mu_j^{\mathrm{neq}} J_j^{\mathrm{e}-2/3} \mathbf{F}_j^{\mathrm{v}-1} \left(\mathbf{I} - \frac{1}{3} \mathrm{tr}\left(\mathbf{C}_j^{\mathrm{e}}\right) \mathbf{C}_j^{\mathrm{e}-1} \right) \mathbf{F}_j^{\mathrm{v}-T} \\ &\quad + \frac{\kappa}{2} (J^2 - 1) \mathbf{C}^{-1}, \\ \eta^{\mathrm{k}} &= c_g \ln\left(\frac{T}{T_0}\right), \\ \eta_i^{\mathrm{c}} &= \Delta c \phi_i \ln \frac{T_{e_i}}{T_2} - \frac{\phi_i \mu^{\mathrm{eq}}}{2T_0} \left(\mathrm{tr}\left(\overline{\mathbf{C}}\right) - 3 \right). \end{aligned} \tag{3.33}$$

The Cauchy stress can then be obtained through $\boldsymbol{\sigma} = \frac{1}{J}\mathbf{FSF}^{\mathrm{T}}$ as

$$\begin{aligned} \boldsymbol{\sigma} &= \frac{1}{J} \sum_i^P \phi_i \frac{T_{e_i}}{T_0} \mu^{\mathrm{eq}} \left(\overline{\mathbf{b}} - \frac{1}{3} \mathrm{tr}\left(\overline{\mathbf{b}}\right) \mathbf{I} \right) + \frac{1}{J} \sum_j^N \mu_j^{\mathrm{neq}} \left(\overline{\mathbf{b}}_j^{e} - \frac{1}{3} \mathrm{tr}\left(\overline{\mathbf{b}}_j^{e}\right) \mathbf{I} \right) \\ &\quad + \frac{1}{2J} \kappa \left(J^2 - 1\right) \mathbf{I}. \end{aligned} \tag{3.34}$$

The Mandel stresses $\mathbf{M}_j^{\mathrm{neq}}$ and $\mathbf{M}_j^{\mathrm{back}}$ can be further obtained as

$$\begin{aligned} \mathbf{M}_j^{\mathrm{neq}} &= \mu_j^{\mathrm{neq}} \left(\overline{\mathbf{C}}_j^{\mathrm{e}} - \frac{1}{3} \mathrm{tr}\left(\overline{\mathbf{C}}_j^{\mathrm{e}}\right) \mathbf{I} \right), \\ \mathbf{M}_j^{\mathrm{back}} &= \frac{\mu_j^{\mathrm{ori}}}{3} \frac{\lambda_{\mathrm{L}}}{\overline{\lambda}_j^{\mathrm{eff}}} \mathcal{L}^{-1} \left(\frac{\overline{\lambda}_j^{\mathrm{eff}}}{\lambda_{\mathrm{L}}} \right) \left(\overline{\mathbf{A}}_j - \frac{1}{3} \mathrm{tr}\left(\overline{\mathbf{A}}_j\right) \mathbf{I} \right). \end{aligned} \tag{3.35}$$

The evolution of $\mathbf{D}_j^{\mathrm{v}}$ is described by Eq. (3.29). The viscosity depends on temperature, structure, and the flow stress. Here, the Adam–Gibbs model [43] is used to represent the dependence of viscosity on temperature and the structural state, while the Eyring model [44] is adopted to describe the stress-activated viscous flow, written as

$$\upsilon_j = \upsilon_j^{\mathrm{ref}} \exp\left(\frac{B}{T \sum_i^P \eta_i^{\mathrm{c}}} \right) \frac{V_S s}{T} \left[\sinh\left(\frac{V_S s}{T} \right) \right]^{-1}, \tag{3.36}$$

where $\upsilon_j^{\mathrm{ref}}$ is the characteristic viscosity at the reference state, B is the activation energy of Adam–Gibbs model, V_S is the activation volume of the Eyring model, and $s = \sqrt{\frac{1}{2} \sum_j^N \mathbf{M}_j^{\mathrm{eff}} : \sum_j^N \mathbf{M}_j^{\mathrm{eff}}}$ is the flow stress. The corresponding stress relaxation time and characteristic stress relaxation time can be defined as $\tau_{S_j} = \upsilon_j / \mu_j^{\mathrm{neq}}$ and $\tau_{S_j}^{\mathrm{ref}} = \upsilon_j^{\mathrm{ref}} / \mu_j^{\mathrm{neq}}$.

Combining Eqs. (3.19), (3.30), and (3.32) leads to the following evolution equation for the orientation tensor $\mathbf{A}_j$,

$$\dot{\mathbf{A}}_j = \mathbf{A}_j \mathbf{D}_j^{\mathrm{v}} + \mathbf{D}_j^{\mathrm{v}} \mathbf{A}_j - \frac{\mathbf{M}_j^{\mathrm{back}}}{\upsilon_j^{\mathrm{ori}}} \mathbf{A}_j. \tag{3.37}$$

The viscosity-type parameter $\upsilon_j^{\rm ori}$ needs to be determined. The hybrid function in Xiao and Tian [39] with a Williams–Landel–Ferry (WLF) function above $T_{\rm g}$ and an Arrhenius function below $T_{\rm g}$ is used to describe the effects of temperature on $\upsilon_j^{\rm ori}$. Similar to the viscosity function, a stress accelerated term is also incorporated into $\upsilon_j^{\rm ori}$. Thus, the following expression is used for $\upsilon_j^{\rm ori}$,

$$\begin{aligned}
\upsilon_j^{\rm ori} &= \upsilon_j^{\rm gori} a(T) \frac{V_s^{\rm ori} s^{\rm ori}}{T} \left[\sinh\left(\frac{V_s^{\rm ori} s^{\rm ori}}{T}\right)\right]^{-1}, \\
\log(a(T)) &= \frac{-C_1^{\rm g}(T-T_0)}{C_2^{\rm g}+T-T_0}, \quad T \geq T_{\rm g}, \\
\log(a(T)) &= -\frac{A}{T}+\frac{A}{T_{\rm g}}, \quad T < T_{\rm g},
\end{aligned} \tag{3.38}$$

where $\upsilon_j^{\rm gori}$ is the characteristic viscosity for the relaxation of the orientation tensor at $T_{\rm g}$, $V_s^{\rm ori}$ is the activation volume parameter, $C_1^{\rm g}$ and $C_2^{\rm g}$ are the WLF constants, $s^{\rm ori} = \sqrt{\frac{1}{2}\sum_j^N \mathbf{M}_j^{\rm back} : \sum_j^N \mathbf{M}_j^{\rm back}}$, and A is the parameter of the Arrhenius function.

Here, we require that the partition parameter a to be a constant value. Thus, the dissipative energy flowing into the kinetic and configurational subsystems is a constant fraction of the total dissipative energy. Thus, by adopting Eqs. (3.29) and (3.30), the requirement for nonnegative values of $\mathcal{W}^{{\rm int},k}$ and $\mathcal{W}_i^{{\rm int},c}$ is satisfied. The total dissipative energy can then be represented as

$$\mathcal{W}^{\rm int} = \sum_j^N \left(\mathbf{M}_j^{\rm eff} : \mathbf{D}_j^{\rm v} + \frac{1}{2\upsilon_j^{\rm ori}} \mathbf{M}_j^{\rm back} : \mathbf{M}_j^{\rm back}\right). \tag{3.39}$$

The evolution equation of the effective temperature is obtained through combining Eqs. (3.26), (3.27), (3.33), (3.39), and (3.23), which gives,

$$\Delta c \phi_i \dot{T}_{e_i} = \kappa_i^{\rm c} \nabla_{\mathbf{X}} \cdot \left(\mathbf{C}^{-1} \nabla_{\mathbf{X}} T_{e_i}\right) + K_i\left(T - T_{e_i}\right) + a\phi_i \mathcal{W}^{\rm int} + \phi_i \mathcal{H}_i^{\rm int}, \quad i = 1 \ldots P, \tag{3.40}$$

where $\mathcal{H}_i^{\rm int} = \frac{J^{-2/3} T_{e_i} \mu^{\rm eq}}{2T_0}\left(\mathbf{I} - \frac{1}{3}{\rm tr}(\mathbf{C})\,\mathbf{C}^{-1}\right) : \dot{\mathbf{C}}$ is a latent heat term.

Using the same set of equations, the first law of the kinetic subsystem can be rewritten as

$$c_g \dot{T} = \kappa^{\rm k} \nabla_{\mathbf{X}} \cdot \left(\mathbf{C}^{-1} \nabla_{\mathbf{X}} T\right) - \sum_i^P K_i\left(T - T_{e_i}\right) + (1-a)\,\mathcal{W}^{\rm int}. \tag{3.41}$$

The first law of the whole system can be rewritten by summing Eqs. (3.40) and (3.41) as

$$\begin{aligned}
c_g \dot{T} + \Delta c \sum_i^P \phi_i \dot{T}_{e_i} &= \kappa^{\rm k} \nabla_{\mathbf{X}} \cdot \left(\mathbf{C}^{-1} \nabla_{\mathbf{X}} T\right) + \sum_i^P \kappa_i^{\rm c} \nabla_{\mathbf{X}} \cdot \left(\mathbf{C}^{-1} \nabla_{\mathbf{X}} T_{e_i}\right) \\
&\quad + \mathcal{W}^{\rm int} + \mathcal{H}^{\rm int}.
\end{aligned} \tag{3.42}$$

We further neglect the heat conduction term in the configurational subsystems. Equation (3.40) is then reduced to

$$\dot{T}_{e_i} = \frac{T - T_{e_i}}{\tau_{R_i}} + \frac{a}{\Delta c} \mathcal{W}^{\rm int} + \frac{1}{\Delta c} \mathcal{H}_i^{\rm int}, \quad i = 1 \ldots P, \tag{3.43}$$

where $\tau_{R_i} = (\Delta c\phi_i)/K_i$ is the structural relaxation time. Without mechanical deformation, the preceding equation reduces to the fictive temperature theory proposed by Tool [45], which is used to describe the volumetric and enthalpic relaxation in amorphous solids [46]. Thus, the first term of the right side of the equation describes the effects caused by structural relaxation, while the second term indicates that the dissipative energy can increase the effective temperature and induce a mechanical rejuvenation effect. The third term is a latent heat term, which is a positive value during the loading process and a negative value during the unloading process. When amorphous solids are deformed in the glassy state, the latent heat is typically much smaller than the viscous dissipative energy.

The same Adam–Gibbs model is used to represent the explicit form for structural relaxation time

$$\tau_{R_i} = \tau_{R_i}^{\text{ref}} \exp\left(\frac{B}{T\sum_i^P \eta_i^{\text{c}}}\right), \tag{3.44}$$

where $\tau_{R_i}^{\text{ref}}$ is the characteristic structural relaxation time in the reference state.

In summary, Eqs. (3.33)–(3.35) define the explicit expression for stress, kinetic entropy and configurational entropy. The evolution equations of viscous deformation velocity $\mathbf{D}_j^{\text{v}}$ and the orientation tensor $\mathbf{A}_j$ are provided in Eqs. (3.29) and (3.37), respectively. Equation (3.43) gives the evolution equation for the effective temperature. The model contains three relaxation mechanisms: the relaxation of viscous deformation, the relaxation of chain orientation, and structural relaxation. Each relaxation is characterized by the corresponding viscosity or relaxation time, with the explicit forms provided in Eqs. (3.36), (3.38), and (3.44).

3.2.4 A Reduced Version of the Constitutive Model

In this part, we simplify the above effective temperature theory model to the viscoelastic case. First, the model does not have the thermodynamic variable, effective temperature, which means that the nonequilibrium structure is always in the equilibrium state. We further do not consider the viscoplastic effect, namely, without the Eyring function for stress-activated viscous flow. The related strong strain hardening effect in the glassy state is also not considered in the work. In addition, we do not consider the mechanical process. Other processes, such as heat conduction, are also not included in the simplified model.

Based on the above assumptions, the Helmholtz free energy density can be written as

$$\Psi = \frac{\mu^{\text{eq}}}{2}\left(\text{tr}\left(\overline{\mathbf{C}}\right) - 3\right) + \sum_j^N \frac{\mu_j^{\text{neq}}}{2}\left(\text{tr}\left(\overline{\mathbf{C}}_j^{\text{e}}\right) - 3\right) + \frac{\kappa}{4}\left(J^2 - 2\log J - 1\right). \tag{3.45}$$

In this condition, the Cauchy stress can be written as

$$\boldsymbol{\sigma} = \frac{1}{J}\mu^{\text{eq}}\left(\overline{\mathbf{b}} - \frac{1}{3}\text{tr}\left(\overline{\mathbf{b}}\right)\mathbf{I}\right) + \frac{1}{J}\sum_j^N \mu_j^{\text{neq}}\left(\overline{\mathbf{b}}_j^e - \frac{1}{3}\text{tr}\left(\overline{\mathbf{b}}_j^e\right)\mathbf{I}\right) + \frac{1}{2J}\kappa\left(J^2 - 1\right)\mathbf{I}. \tag{3.46}$$

The evolution equation of $\mathbf{D}_j^{\text{v}}$ is written as

$$\mathbf{D}_j^{\text{v}} = \frac{\mathbf{M}_j}{2v_j}, \quad \mathbf{M}_j = \mu_j^{\text{neq}}\left(\overline{\mathbf{C}}_j^{\text{e}} - \frac{1}{3}\text{tr}\left(\overline{\mathbf{C}}_j^{\text{e}}\right)\mathbf{I}\right), \quad j = 1 \ldots N, \tag{3.47}$$

where $\mathbf{M}_j$ is the Mandel stress. Again, the stress relaxation time is still related to viscosity as $\tau_{S_j} = \upsilon_j/\mu_j^{\text{neq}}$.

The above model is in fact the same as the first finite deformation viscoelastic model introduced in Chapter 2. However, to model the shape-memory effect, the relaxation time needs to be dependent on the temperature. Here, we adopt the WLF equation for the stress relaxation time in the glass transition region and the Arrhenius function below the glass transition temperature as

$$\begin{aligned} &\tau_{S_j} = \tau_{S_j}^{\text{ref}} a(T), \\ &\log(a(T)) = \frac{-C_1^{\text{g}}(T - T_0)}{C_2^{\text{g}} + T - T_0}, \quad T \geq T_{\text{g}}, \\ &\log(a(T)) = -\frac{A}{T} + \frac{A}{T_{\text{g}}}, \quad T < T_{\text{g}}, \end{aligned} \tag{3.48}$$

It should be noted that in the above equation the Adam–Gibbs model is adopted for the stress relaxation time. The Adam–Gibbs model in the glass transition region can be equivalent to the WLF function through choosing a specific form for the heat capacity.

3.3 Parameter Determination

For the simplified viscoelastic model, the parameters can be determined from a single set of rheological tests [32, 47]. These tests can be performed in the form of frequency sweep tests or stress relaxation tests at multiple temperatures. A master curve can then be constructed following the time-temperature superposition. The shift factor can be employed to determine the WLF constants. For the effective temperature theory, additional parameters related to structural relaxation can be obtained from differential scanning calorimetry [34, 48]. Finally, the parameters related to the viscoplastic behaviors can be fitted to the stress–strain responses at different strain rates below T_{g} [34, 39]. Here, we list the related parameters for an amorphous polymer, poly(ethylene terephthalate)-glycol (PETG), in Tables 3.1 and 3.2, while the detailed procedures can be found in Dai et al. [34].

3.4 Results

3.4.1 Performance of the Viscoelastic Model

We first apply the finite deformation viscoelastic model to simulate the shape-memory behaviors. The parameters related to the viscoelastic model are the discrete stress relaxation spectrum ($\tau_{S_j}^{\text{g}} - \mu_j^{\text{g}}$) listed in Table 3.2, the WLF constants C_1^{g}, C_2^{g}, the Arrhenius parameter A and the bulk modulus κ shown in Table 3.1. We first show the glass transition behavior predicted by the model through plotting the rheological properties. Specifically, we show the shear storage modulus, calculated as

$$\mu' = \mu^{\text{eq}} + \sum_{j}^{N} \mu_j^{\text{neq}} \frac{\omega^2 \tau_{S_j}^2(T)}{1 + \omega^2 \tau_{S_j}^2(T)}, \tag{3.49}$$

Table 3.1 Parameters of the constitutive model for the amorphous polymer PETG.

Parameter	Values	Physical significance
μ^{eq} (MPa)	0.93	Equilibrium shear modulus
κ (MPa)	1166.67	Bulk modulus
$B(J/g)$	240.2	Thermal activation energy
$T_2(K)$	323.5	Kauzmann temperature
$T_g(K)$	345	Glass transition temperature
$T_0(K)$	375	Reference temperature
C_1^g	11.78	The first WLF constant
$C_2^g(K)$	21.47	The second WLF constant
$A(K)$	15 000	The Arrhenius parameter for the relaxation below T_g
$V_s(K/MPa)$	125	Activation parameter for viscous flow
c_g (J/(gK))	1.09	Heat capacity of kinetic subsystem
Δc (J/(gK))	0.42	Excess heat capacity of configurational subsystems
a	0.67	Partition parameter
λ_L	1.7	Stretch limit of the eight-chain model
μ_j^{ori}/μ_j^{neq}	0.0114	The ratio between shear modulus μ_j^{ori} and μ_j^{neq}
$\tau_j^{gori}/\tau_{S_j}^{g}$	100.0	The ratio between relaxation time τ_j^{gori} and τ_j^{g}
$V_s^{ori}(K/MPa)$	150.0	Activation parameter for relaxation of network orientation

where ω is the applied angular frequency. The shear loss modulus can be obtained as

$$\mu'' = \sum_j^N \mu_j^{neq} \frac{\omega \tau_{S_j}(T)}{1 + \omega^2 \tau_{S_j}^2(T)}. \tag{3.50}$$

The corresponding loss factor is defined as

$$\tan \delta = \frac{\mu''}{\mu'}. \tag{3.51}$$

As shown in Figure 3.1, PETG exhibits a glass transition region from 350 to 370 K, which is relatively narrow. The peak of loss factor occurs at around 360 K.

The viscoelastic model can be employed to simulate the shape-memory behaviors of polymers programmed above T_g. We first investigate the effect of temperature on the recovery response. A constant strain rate of 0.003/s is applied to obtain a 30% engineering strain at the deformation temperature, 360 or 350 K. The temperature is then decreased to 320 K at 3 °C/min. The force is further reduced to zero. During the recovery process, the temperature is increased to 380 K at 3 °C/min. As shown in Figure 3.2a, the shape recovery occurs in a lower temperature region for a lower deformation temperature. This is the so-called temperature-memory effect. However, compared to other amorphous polymer systems with a broad glass transition region [47], the temperature memory effect in PETG is

Table 3.2 Discrete stress and structural relaxation spectra for PETG.

Number	$\tau^{g}_{S_j}$ (sec)	μ^{neq}_{j} (MPa)	$\tau^{g}_{R_i}$ (sec)	ϕ_i
1	0.0186	28.54	0.003	0.0123
2	0.073	26.30	0.0125	0.00739
3	0.286	43.51	0.0524	0.0114
4	1.116	34.02	0.219	0.0176
5	4.383	37.36	0.917	0.0269
6	17.15	56.41	3.836	0.0408
7	67.26	68.02	16.04	0.0606
8	263.1	53.93	67.08	0.0877
9	1031	27.43	280.52	0.1209
10	4038.9	9.461	1173.05	0.1541
11	15817	2.5812	4905.36	0.1718
12	61941	0.6896	20512.86	0.1528
13	242575	0.2385	85779.06	0.0946
14	949958	0.2023	358704.03	0.0338
15	3720080	0.1858	1500000	0.0071

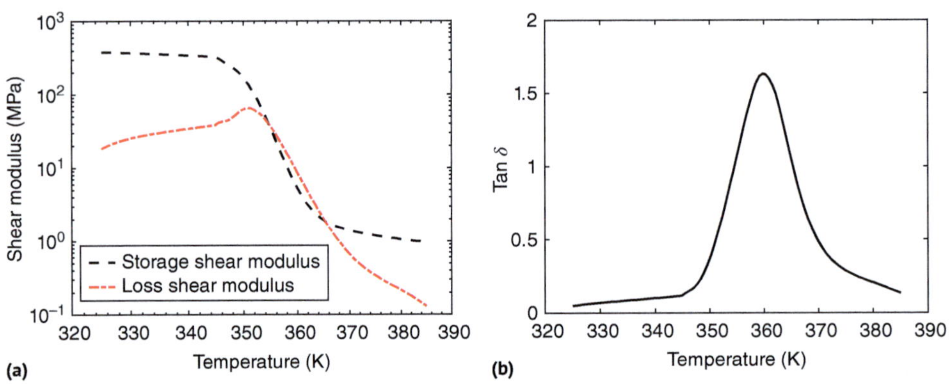

Figure 3.1 The rheological behavior predicted by the model: (a) storage shear modulus and loss shear modulus and (b) loss factor.

relatively weak. We also apply the model to study the effect of heating rate. The deformation temperature is set as 360 K, while the heating rate is chosen as 1 °C/min and 5 °C/min. As shown in Figure 3.2b, the model predicts an increase in recovery temperature with a higher heating rate, which is consistent with the experimental observations [15]. The effect of heating rate is mainly related to the stress relaxation. A faster heating rate leads to less time for relaxation. Thus, a higher temperature is needed to reach the same level of shape recovery.

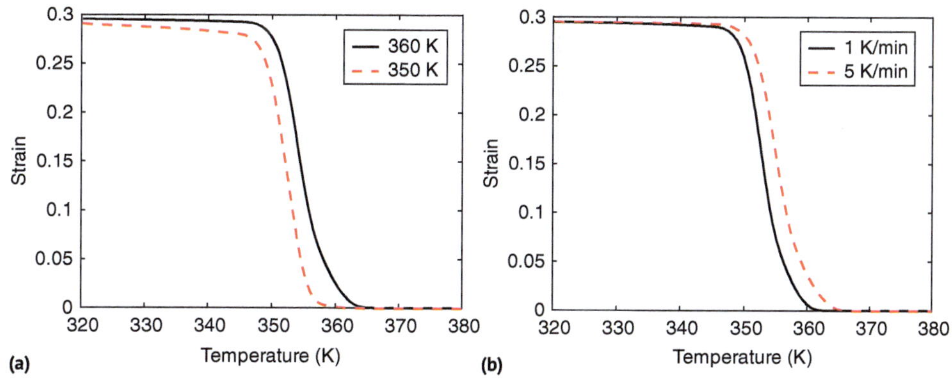

Figure 3.2 The shape recovery performance predicted by the viscoelastic model subjected to: (a) different deformation temperatures and (b) different heating rates.

3.4.2 Performance of the Effective Temperature Model

Compared with the viscoelastic model, two additional mechanisms have been incorporated into the effective temperature theory, i.e. the structural relaxation and stress-activated viscoplastic flow. If the shape-memory effect is not strongly related to the above two mechanisms, such as the case shown in the above part, the predicted responses from the viscoelastic model and the effective temperature model are almost the same. However, the effective temperature theory can also predict other phenomena, which is beyond the ability of the viscoelastic model. First, we investigate the aging effect. The programming process is set the same as the previous cases. A constant strain rate of 0.003/s is applied to obtain a 30% engineering strain at the deformation temperature of 360 K. The temperature is then decreased to 320 K at 3 °C/min. The force is also reduced to zero. A different annealing time is then applied at 320 K before increasing the temperature to 380 K at 3 °C/min. Figure 3.3 compares the response without annealing and with an annealing time of 14 days. The results show that the aging treatment induces a higher initial temperature and a much larger initial recovery rate. These findings are consistent with the experimental observations reported in Choi et al. [18]. During the aging process, the nonequilibrium structure continuously evolves toward equilibrium, represented as a decrease in effective temperature in the model. The viscosity is related to the structure state. A more sluggish structure results in a higher viscosity, namely, less shape recovery within the same period of time. During heating, the structure is rejuvenated by the thermal process. However, a higher temperature is required to rejuvenate more sluggish structures. This leads to a higher recovery temperature. Once rejuvenated, the frozen shape can be quickly activated, resulting in an initial faster recovery rate.

The effective temperature theory also incorporates the viscoplastic deformation mechanism. Thus, it can describe the stress response of amorphous polymers across the glass transition region [40]. As shown in Figure 3.4, the model reproduces the main features of the stress response of amorphous polymers deformed below T_g at an engineering strain rate of 0.0003/s, including the initial linear region, the strain softening region, the steady state

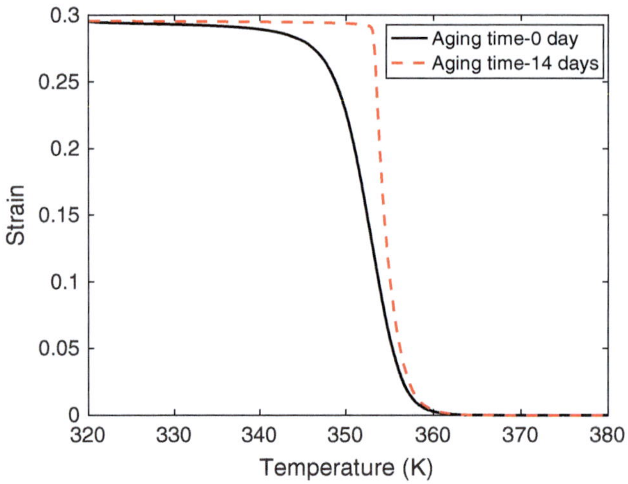

Figure 3.3 The effect of aging time on the shape recovery performance of PETG predicted by the effective temperature theory.

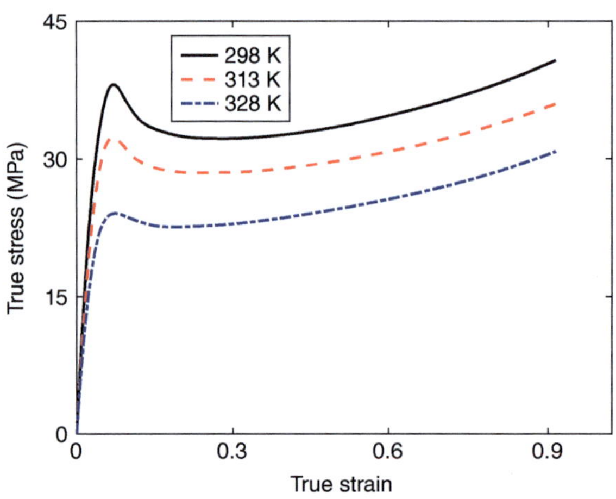

Figure 3.4 The stress response of PETG at different temperatures predicted by the effective temperature theory.

region, and the hardening region. The strain softening is caused by the mechanical rejuvenation. The viscoplastic dissipation flows into the configurational subsystems, resulting in an increase in the effective temperature and a further decrease in viscosity. The strain hardening is caused by the orientation of polymer networks. In Dai et al. [34], we have shown that this model can effectively capture the rate- and temperature-dependent stress responses. The model can also well predict the stress response of amorphous polymers subjected to different aging conditions, as shown in Xiao and Nguyen [40].

To evaluate shape-memory performance, either the free recovery or the constrained recovery tests can be used. In the constrained recovery tests, the strain is fixed, while the stress is measured during the heating process. Previous works [7, 14] have shown that the recovery stress is only several megapascals. It is found that cold programming can effectively increase the recovery stress [7]. Here, we apply the effective temperature theory to

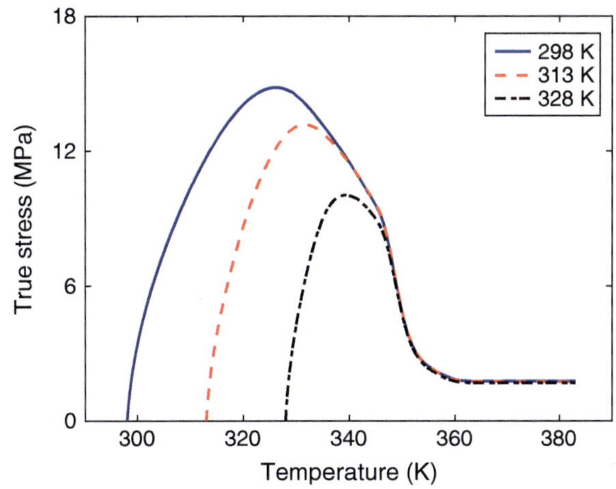

Figure 3.5 The recovery stress as a function of temperature for different programming temperatures in the constrained recovery tests predicted by the effective temperature theory.

investigate the constrained recovery tests. The programming process involves applying a 60% engineering strain at an engineering strain rate of 0.0003/s below T_g and then cooling to 293 K. The force is then unloaded. The heating rate is chosen as 0.5 °C/min. Figure 3.5 shows that the recovery stress first increases with temperature until reaching the peak value and then decreases with temperature to the final plateau value. Both the recovery regions and the peak stress depend on the programming temperature. A lower programming temperature leads to a lower activation temperature and a higher peak stress. It is noted that the peak stress can reach as high as 15 MPa, much higher than the magnitude obtained for the hot programming conditions.

3.5 Discussion

The thermally activated process achieves the shape recovery through increasing the temperature. Alternatively, the programmed shape can be activated through exposure to moisture and solvent environment in a thermal condition [4]. This is because solvent molecules can diffuse into the polymer matrix to increase the chain mobility, i.e. a decrease in viscosity. The effective temperature theory can be extended to simulate the solvent-driven shape-memory effect. First, the relaxation time should have a solvent dependence, which can be modified by assuming a change in the configurational entropy when the solvent is introduced as

$$\eta^c = \eta^c_{\text{dry}} + \Delta\eta^c, \tag{3.52}$$

where $\Delta\eta^c$ is the change in the configurational entropy. Even a small fraction of solvent can induce a much softer stress response as demonstrated in Figure 3.6. The explicit form for $\Delta\eta^c$ and the related parameters can refer to Xiao and Nguyen [26].

The solvent gradually diffuses into the polymer matrix. Thus, it is also necessary to simulate the diffusion process. This can be achieved by adopting the chemo-mechanical framework for modeling diffusion in gels [28, 49]. Alternatively, Fick's type model can also be employed [26]. In this case, the diffusion process has the same form as the heat conduction

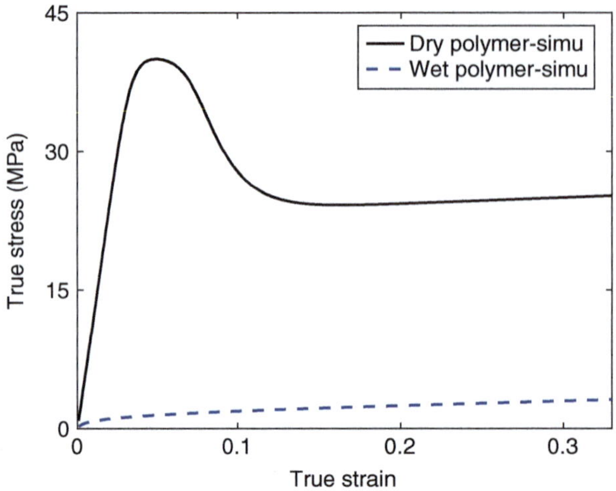

Figure 3.6 The predicted stress response of dry and wet polymers. The detailed information for the model and parameters can be found in Xiao and Nguyen [26]. *Source*: Adapted from Xiao and Nguyen (2013).

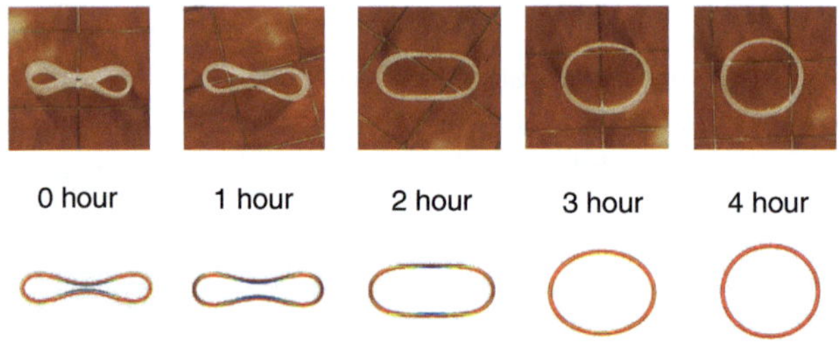

Figure 3.7 Comparison between the measured and predicted shape recovery of a stent in water at 30 °C. *Source*: Reproduced with permission from Xiao and Nguyen (2013)/The Royal Society of Chemistry.

process described by the Fourier's law. Xiao and Nguyen [26] performed experiments to investigate the recovery of an SMP stent in water at 30 °C, where the dry polymers are in a glassy state. As shown in Figure 3.7, the stent can recover to its original shape in four hours. Through incorporating the effect of solvents on relaxation time and the diffusion process, the model can accurately predict the whole recovery process. It should be noted that the solvent can also change the size of the SMP due to the swelling effect, which should also be incorporated into the model.

3.6 Conclusion

In this work, we presented an effective temperature theory for describing the shape-memory effect in amorphous polymers. This framework introduces effective temperature and configurational entropy as a pair of thermodynamic conjugate variables to describe the nonequilibrium structural state. The model reduces to the classic glassy–rubbery transition model when the structure is in equilibrium. Ignoring the stress-activated viscous flow, the model can be further simplified as the finite deformation viscoelastic model. Thus, the effective temperature theory can capture almost all the main features of thermomechanical behaviors and shape-memory performances of amorphous networks. It can well describe the stress response across the glass transition regime. It can also predict the aging and mechanical predeformation on the stress response. It also captures the shape-memory effect under various programming and recovery conditions, such as deformation temperatures, aging time, heating rate, etc. We also show the possibility of extending the effective temperature theory for shape recovery in solvent and moisture conditions.

References

1 Yuliang Xia, Yang He, Fenghua Zhang, Yanju Liu, and Jinsong Leng. A review of shape memory polymers and composites: mechanisms, materials, and applications. *Advanced Materials*, 33(6): 2000713, 2021.

2 Jinsong Leng, Xin Lan, Yanju Liu, and Shanyi Du. Shape-memory polymers and their composites: stimulus methods and applications. *Progress in Materials Science*, 56(7): 1077–1135, 2011.

3 Changhong Linghu, Tong Mu, Wei Zhao, Yangchengyi Liu, K Jimmy Hsia, Jinsong Leng, and Huajian Gao. Advancing smart dry adhesives with shape memory polymers. *International Journal of Smart and Nano Materials*, 16(1): 103–143, 2025.

4 Rui Xiao and Wei Min Huang. Heating/solvent responsive shape-memory polymers for implant biomedical devices in minimally invasive surgery: current status and challenge. *Macromolecular Bioscience*, 20(8): 2000108, 2020.

5 Frank Katzenberg and Joerg C Tiller. Shape memory natural rubber. *Journal of Polymer Science Part B: Polymer Physics*, 54(14): 1381–1388, 2016.

6 Nasim Sabahi, Iman Roohani, Chun H Wang, Ehsan Farajzadeh, and Xiaopeng Li. Thermoplastic polyurethane-based shape memory polymers with potential biomedical application: the effect of TPU soft-segment on shape memory effect and cytocompatibility. *Polymer*, 283: 126189, 2023.

7 Guoqiang Li and Anqi Wang. Cold, warm, and hot programming of shape memory polymers. *Journal of Polymer Science Part B: Polymer Physics*, 54(14): 1319–1339, 2016.

8 Tao Xie, Kiart A Page, and Scott A Eastman. Strain-based temperature memory effect for nafion and its molecular origins. *Advanced Functional Materials*, 21: 2057–2066, 2011.

9 Nikolaus Mirtschin and Thorsten Pretsch. Designing temperature-memory effects in semicrystalline polyurethane. *RSC Advances*, 5(57): 46307–46315, 2015.

10 Alicia M Ortega, Christopher M Yakacki, Sean A Dixon, Roxanne Likos, Alan R Greenberg, and Ken Gall. Effect of crosslinking and long-term storage on the shape-memory behavior of (meth) acrylate-based shape-memory polymers. *Soft Matter*, 8(28): 7381–7392, 2012.

11 Tao Xie. Tunable polymer multi-shape memory effect. *Nature*, 464(7286): 267–270, 2010.

12 Mohadeseh Zare, Molamma P Prabhakaran, Nader Parvin, and Seeram Ramakrishna. Thermally-induced two-way shape memory polymers: mechanisms, structures, and applications. *Chemical Engineering Journal*, 374: 706–720, 2019.

13 Thao D Nguyen. Modeling shape-memory behavior of polymers. *Polymer Reviews*, 53(1): 130–152, 2013.

14 Thao D Nguyen, H Jerry Qi, Francisco Castro, and Kevin N Long. A thermoviscoelastic model for amorphous shape memory polymers: incorporating structural and stress relaxation. *Journal of the Mechanics and Physics of Solids*, 56: 2792–2814, 2008.

15 Kai Yu and H Jerry Qi. Temperature memory effect in amorphous shape memory polymers. *Soft Matter*, 10(47): 9423–9432, 2014.

16 Julie Diani, Pierre Gilormini, Carole Frédy, and Ingrid Rousseau. Predicting thermal shape memory of crosslinked polymer networks from linear viscoelasticity. *International Journal of Solids and Structures*, 49(5): 793–799, 2012.

17 Vikas Srivastava, Shawn A Chester, and Lallit Anand. Thermally actuated shape-memory polymers: experiments, theory, and numerical simulations. *Journal of the Mechanics and Physics of Solids*, 58(8): 1100–1124, 2010.

18 Jinwoo Choi, Alicia M Ortega, Rui Xiao, Christopher M Yakacki, and Thao D Nguyen. Effect of physical aging on the shape-memory behavior of amorphous networks. *Polymer*, 53(12): 2453–2464, 2012.

19 Ming Lei, Kai Yu, Haibao Lu, and H Jerry Qi. Influence of structural relaxation on thermomechanical and shape memory performances of amorphous polymers. *Polymer*, 109: 216–228, 2017.

20 H Jerry Qi, Thao D Nguyen, Francisco Castro, Christopher Yakacki, and Robin Shandas. Finite deformation thermo-mechanical behavior of thermally-induced shape memory polymers. *Journal of the Mechanics and Physics of Solids*, 56: 1730–1751, 2008.

21 Yiheng Xue, Jincheng Lei, and Zishun Liu. A thermodynamic constitutive model for shape memory polymers based on phase transition. *Polymer*, 243: 124623, 2022.

22 Hao Zeng, Huiyu Sun, and Jianpin Gu. Modeling the one-way and two-way shape memory effects of semi-crystalline polymers. *Smart Materials and Structures*, 30(9): 095020, 2021.

23 Jianping Gu, Changchun Wang, Hao Zeng, Hao Duan, Mengqi Wan, and Huiyu Sun. A thermo-mechanical constitutive model for triple-shape and two-way shape memory polymers. *Smart Materials and Structures*, 33(6): 065034, 2024.

24 Matteo Arricca, Nicoletta Inverardi, Stefano Pandini, Maurizio Toselli, Massimo Messori, Ferdinando Auricchio, and Giulia Scalet. Phenomenological modeling of the stress-free two-way shape-memory effect in semi-crystalline networks: formulation, numerical simulation, and experimental validation. *European Journal of Mechanics-A/Solids*, 105: 105245, 2024.

25 Hasan Gülaşik, Maxime Houbben, Clara Pereira Sànchez, Juan Manuel Calleja Vázquez, Philippe Vanderbemden, Christine Jérôme, and Ludovic Noels. A thermo-mechanical, viscoelasto-plastic model for semi-crystalline polymers exhibiting one-way and two-way shape memory effects under phase change. *International Journal of Solids and Structures*, 295: 112814, 2024.

26 Rui Xiao and Thao D Nguyen. Modeling the solvent-induced shape-memory behavior of glassy polymers. *Soft Matter*, 9(39): 9455–9464, 2013.

27 Yiqi Mao, Fen Chen, Shujuan Hou, H Jerry Qi, and Kai Yu. A viscoelastic model for hydrothermally activated malleable covalent network polymer and its application in shape memory analysis. *Journal of the Mechanics and Physics of Solids*, 127: 239–265, 2019.

28 Lu Dai and Rui Xiao. A thermodynamic-consistent model for the thermo-chemo-mechanical couplings in amorphous shape-memory polymers. *International Journal of Applied Mechanics*, 13(02): 2150022, 2021.

29 Madhubhashitha Herath, Jayantha Epaarachchi, Mainul Islam, Liang Fang, and Jinsong Leng. Light activated shape memory polymers and composites: a review. *European Polymer Journal*, 136: 109912, 2020.

30 Kevin N Long, Timothy F Scott, H Jerry Qi, Christopher N Bowman, and Martin L Dunn. Photomechanics of light-activated polymers. *Journal of the Mechanics and Physics of Solids*, 57(7): 1103–1121, 2009.

31 Rui Xiao, Jinwoo Choi, Nishant Lakhera, Christopher M Yakacki, Carl P Frick, and Thao D Nguyen. Modeling the glass transition of amorphous networks for shape-memory behavior. *Journal of the Mechanics and Physics of Solids*, 61(7): 1612–1635, 2013.

32 Rui Xiao, Christopher M Yakacki, Jingkai Guo, Carl P Frick, and Thao D Nguyen. A predictive parameter for the shape memory behavior of thermoplastic polymers. *Journal of Polymer Science Part B: Polymer Physics*, 54(14): 1405–1414, 2016.

33 Cheng Zhang, Xiaofan Gou, and Rui Xiao. Controllable shape-memory recovery regions in polymers through mechanical programming. *Journal of Applied Polymer Science*, 135(8): 45909, 2018.

34 Lu Dai, Chuanshuai Tian, and Rui Xiao. Modeling the thermo-mechanical behavior and constrained recovery performance of cold-programmed amorphous shape-memory polymers. *International Journal of Plasticity*, 127: 102654, 2020.

35 Th M Nieuwenhuizen. Thermodynamics of the glassy state: effective temperatureas an additional system parameter. *Physical Review Letters*, 80: 5580–5583, 1998.

36 Eran Bouchbinder and J S Langer. Nonequilibrium thermodynamics of driven amorphous materials. II. Effective-temperature theory. *Physical Review E*, 80: 031132, 2009.

37 Ken Kamrin and Eran Bouchbinder. Two-temperature continuum thermomechanics of deforming amorphous solids. *Journal of the Mechanics and Physics of Solids*, 73: 269–288, 2014.

38 Charles O Frederick and Paul J Armstrong. A mathematical representation of the multiaxial bauschinger effect. *Materials at High Temperatures*, 24(1): 1–26, 2007.

39 Rui Xiao and Chuanshuai Tian. A constitutive model for strain hardening behavior of predeformed amorphous polymers: incorporating dissipative dynamics of molecular orientation. *Journal of the Mechanics and Physics of Solids*, 125: 472–487, 2019.

40 Rui Xiao and Thao D Nguyen. An effective temperature theory for the nonequilibrium behavior of amorphous polymers. *Journal of the Mechanics and Physics of Solids*, 82: 62–81, 2015.

41 Gerhard A Holzapfel. *Nonlinear Solid Mechanics: a Continuum Approach for Engineers*. Chichester: John Wiley and Sons, LTD, 2000.

42 Ellen M Arruda and Mary C Boyce. A three-dimensional constitutive model for the large stretch behavior of rubber elastic materials. *Journal of the Mechanics and Physics of Solids*, 41: 389–412, 1993.

43 Gerold Adam and Julian H Gibbs. On the temperature dependence of cooperative relaxation properties in the glass-forming liquids. *The Journal of Chemical Physics*, 43: 139–146, 1965.

44 Henry Eyring. Viscosity, plasticity, and diffusion as examples of absolute reaction rates. *The Journal of Chemical Physics*, 4: 283–291, 1936.

45 Arthur Q Tool. Viscosity and extraordinary heat effects in glass. *Journal of the American Ceramic Society*, 29: 240–253, 1946.

46 Rui Xiao, Jun Guo, Chuanshuai Tian, and Wen Chen. Modeling enthalpy relaxation using the mittag-leffler function. *Journal of Non-Crystalline Solids*, 465: 17–25, 2017b.

47 Rui Xiao, Jingkai Guo, and Thao D Nguyen. Modeling the multiple shape memory effect and temperature memory effect in amorphous polymers. *RSC Advances*, 5(1): 416–423, 2015.

48 Rui Xiao, Gagik Ghazaryan, Theo A Tervoort, and Thao D Nguyen. Modeling energy storage and structural evolution during finite viscoplastic deformation of glassy polymers. *Physical Review E*, 95(6): 063001, 2017a.

49 Wei Hong, Xuanhe Zhao, Jinxiong Zhou, and Zhigang Suo. A theory of coupled diffusion and large deformation in polymeric gels. *Journal of the Mechanics and Physics of Solids*, 56(5): 1779–1793, 2008.

4

An Electromechanical Coupled Model for Dielectric Elastomers

Dielectric elastomers (DEs) can exhibit large deformation upon the electric field, which shows promising applications in various areas. This chapter presents a thermodynamic framework for DEs. An electromechanical coupled model is then developed, which includes the viscoelastic effect. Using the model, we first analyze the electromechanical instability (EMI) of DEs. The chapter implements the model for finite element analysis. The results show that rate-dependent viscoelastic properties considerably affect the actuation performance. The framework presented in this chapter can be effectively employed to design actuators based on DEs.

4.1 Introduction

Dielectric elastomers (DEs) are soft electroactive polymers, which can generate large deformation upon an electric field. Due to fast response characteristics [1, 2], DEs can be widely applied in soft actuators, sensors, artificial muscles, soft robotics, etc. [3–9]. For example, Shian et al. [10] developed a soft gripper integrating stiff fibers with DEs, which enable the gripper with flexible deformation to grasp different shapes. Feng et al. [8] reported an ultra-fast soft robot based on DEs, while Liu et al. [11] developed self-powered and self-healable artificial muscles for long-term usage based on DEs. To improve the performance, DEs are supposed to exhibit soft elasticity, low viscosity, high electrical breakdown strength, and high dielectric constant [12]. In the past decades, various DE materials have been synthesized to optimize their properties. The most widely used DEs are the acrylic-based elastomers. Several other types such as silicon-based and polyurethane-based elastomers have also been developed in recent years [13–16].

It is highly demanding to develop electromechanical coupling theories for DEs, which have garnered significant research attention over the years [17–26]. McMeeking and Landis [17] developed a theoretical framework for DEs in the deformed state, leveraging the true electric field and the true electric displacement as fundamental quantities. Suo et al. [18] formulated a nonlinear theory that utilizes thermodynamic conjugate variables, namely, the nominal electric field and the nominal electric displacement, to describe the behavior of these materials. Henann et al. [27] constructed a thermodynamic framework for electromechanical coupling behaviors in DEs. These models provide a complete description of

Multi-field Modeling of Soft Active Materials: Properties and Design, First Edition. Rui Xiao.

electroactivated behaviors in DEs and also demonstrate potentials in designing actuators and energy-harvesting devices based on DEs. Mehnert et al. [28] extended this research by developing a thermo-electro-mechanical coupled theory capable of predicting the response of DEs under varying thermal, electrical, and mechanical loading conditions. Chen et al. [23] established a visco-hyperelastic-plastic model to capture the cyclic deformation behavior of DEs under complex combinations of mechanical and electric loads, underscoring the versatility and adaptability of these theories for diverse applications.

In the realm of both experimental and theoretical investigations, the majority of current research endeavors focus on acrylic-based DEs, particularly VHB elastomers. These materials exhibit a pronounced plateau region within their stress–strain relationship. Upon the application of an electric field that surpasses a critical threshold, these elastomers undergo a dramatic, abrupt deformation phenomenon known as EMI, typically accompanied by subsequent electric breakdown [29–31]. This instability stems from the point where the increasing Maxwell stress surpasses the resistance imparted by the material's mechanical stiffness. To address this issue, multiple strategies have been proposed, with pre-stretching emerging as a highly effective method, as evidenced by both theoretical predictions and experimental validations [32–34]. By inducing sufficient pre-stretching, elastomers can be shifted into a strain-stiffening regime, thereby enhancing their resistance against further thickness reduction. Nevertheless, the implementation of such initial pre-strain typically necessitates the use of a rigid frame, posing a significant constraint for practical applications and hindering widespread adoption.

The main goal of this chapter is to investigate the electromechanical behavior of DE materials, especially the process of obtaining constitutive equations from a nonequilibrium thermodynamic framework. The viscoelastic dissipation will also be incorporated to investigate the rate-dependent activated response of DEs. Further, the equivalent forms of constitutive equations will be provided for implementation of the finite element program COMSOL Multiphysics, which is suitable to simulating the multi-field coupled responses.

This chapter is organized as follows. Section 4.2 introduces a general process to build a nonlinear field constitutive model for coupled multiple fields. A theoretical analysis is then performed to determine the instability of DE materials, based on the energy minimization principle. Section 4.3 simulates several demos of DE actuators using finite element analysis. The effect of viscoelasticity on the electroactivated response is further analyzed.

4.2 Theory

This section presents a general process to construct a constitutive model for DEs. We start with a thermodynamic framework for the electromechanical coupled process. The framework is then simplified for an ideal model with the neo-Hookean model for the mechanical part and the ideal electric energy function. To introduce the viscoelastic effects, the model proposed by Linder et al. [35] is employed.

4.2.1 Thermodynamic Framework

4.2.1.1 Kinematics

The one-to-one mapping between an arbitrary material point $\mathbf{X}$ in the reference configuration Ω_0 and the spatial point $\mathbf{x}$ at time t in the current configuration Ω is given by $\mathbf{x} = \chi(\mathbf{X}, t)$. The deformation gradient is defined as

$$\mathbf{F} = \frac{\partial \mathbf{x}}{\partial \mathbf{X}}. \tag{4.1}$$

The right and left Cauchy–Green tensors are given by $\mathbf{C} = \mathbf{F}^{\mathrm{T}}\mathbf{F}$ and $\mathbf{b} = \mathbf{F}\mathbf{F}^{\mathrm{T}}$, respectively. The volume ratio (or Jacobian determinant) J is calculated as $J = \det \mathbf{F}$.

4.2.1.2 Electric Quantities

The analysis of the electric field's contribution to the dielectric elastomeric system is the essential part. We denote the electric potential as $\varphi(\mathbf{X}, t)$, which is defined in the reference configuration. The reference electric field is defined as

$$\mathbf{E}_{\mathrm{R}} = -\mathrm{Grad}\varphi. \tag{4.2}$$

Following the form of Faraday's law, the reference electric field obeys the rule

$$\mathrm{Curl}\mathbf{E}_{\mathrm{R}} = \mathbf{0}. \tag{4.3}$$

Similarly, we can define the electric field in the current configuration as

$$\mathbf{E}_{\mathrm{c}} = -\mathrm{grad}\varphi. \tag{4.4}$$

Here we use $\mathbf{E}_{\mathrm{c}}$ rather than $\mathbf{E}$ to distinguish it from the Green–Lagrange strain tensor in other chapters. The corresponding electrostatic form of Faraday's law in the current configuration is written as

$$\mathrm{curl}\mathbf{E}_{\mathrm{c}} = \mathbf{0}. \tag{4.5}$$

Since the applied voltage is objective, the integration of reference electric field strength along the reference line element $\mathbf{dL}$ should be equal to the integration of the current electric field strength along the current line element $\mathbf{dl}$, which gives

$$\int \mathbf{E}_{\mathrm{R}}\mathbf{dL} = \int \mathbf{E}_{\mathrm{c}}\mathbf{dl}. \tag{4.6}$$

The relation between two line elements in different configurations is given as

$$\mathbf{F}\mathbf{dL} = \mathbf{dl}. \tag{4.7}$$

Thus, we can obtain the following relationship between the electric field in the reference and deformed configurations:

$$\mathbf{E}_{\mathrm{R}} = \mathbf{F}^{\mathrm{T}}\mathbf{E}_{\mathrm{c}}. \tag{4.8}$$

4.2.1.3 Balance of Momentum

We denote the boundary of the reference body by $\partial\Omega_0$, and the first Piola–Kirchhoff stress tensor is denoted as $\mathbf{P}$. Neglecting the inertial force, the balance of momentum in Ω_0 is expressed as

$$\mathrm{Div}\mathbf{P} + \mathbf{B} = \mathbf{0}, \tag{4.9}$$

where $\mathbf{B}$ is the body force per unit volume in Ω_0.

The external surface traction $\mathbf{T}$ on the boundary $\partial\Omega_0$ is given by

$$\mathbf{T} = \mathbf{PN}, \tag{4.10}$$

where $\mathbf{N}$ is the outward unit vector normal to the boundary $\partial\Omega_0$.

The first Piola–Kirchhoff stress $\mathbf{P}$ is related to the Cauchy stress $\boldsymbol{\sigma}$ in the deformed configuration Ω as

$$\boldsymbol{\sigma} = \frac{1}{J}\mathbf{PF}^{\mathrm{T}}. \tag{4.11}$$

4.2.1.4 Gauss's Law

The electric displacement in the reference configuration is denoted as $\mathbf{D}_{\mathrm{R}}$. The Gauss's law can then be expressed when neglecting the electrodynamic effects as

$$\mathrm{Div}\mathbf{D}_{\mathrm{R}} = \hat{q}, \tag{4.12}$$

where $\hat{q}$ is the free charge density per unit volume in the reference configuration.

The free charge density on a surface boundary follows

$$\hat{\omega} = -\mathbf{D}_{\mathrm{R}}\mathbf{N}. \tag{4.13}$$

For a given area, the electric flux is objective, thus we can have

$$\int \mathbf{D}_{\mathrm{R}}\mathrm{d}\mathbf{S} = \int \mathbf{D}\mathrm{d}\mathbf{s}, \tag{4.14}$$

where $\mathbf{D}$ is the corresponding electric displacement in the deformed configuration, $\mathbf{S}$ and $\mathbf{s}$ are the area element in reference and current configurations, respectively. Nanson's formula is further used to relate the area elements as

$$\mathrm{d}\mathbf{s} = J\mathbf{F}^{-\mathrm{T}}\mathrm{d}\mathbf{S}. \tag{4.15}$$

We can then have

$$\mathbf{D}_{\mathrm{R}} = J\mathbf{F}^{-1}\mathbf{D}. \tag{4.16}$$

4.2.1.5 Energy Imbalance

The free energy imbalance requires that the temporal increase in the Helmholtz free energy of any region is less than or equal to the power expended externally on that region. Letting Ψ denote the Helmholtz free energy per unit volume, thus the imbalance has the form

$$\frac{\mathrm{d}}{\mathrm{d}t}\int_{\Omega_0} \Psi \mathrm{dV} \leq \oint_{\partial\Omega_0} \mathbf{T}\cdot\dot{\mathbf{V}}\mathrm{dS} + \int_{\Omega_0} \mathbf{B}\cdot\dot{\mathbf{V}}\mathrm{dV} + \oint_{\partial\Omega_0} \varphi\cdot\dot{\hat{\omega}}\mathrm{dS} + \int_{\Omega_0} \varphi\cdot\dot{\hat{q}}\mathrm{dV}, \tag{4.17}$$

where **V** is velocity. The first two terms on the right side in the equation represent the mechanical power induced by the traction and body force, and the last two terms represent the electric energy power caused by the free charges on the surface and in the body.

Substituting both the mechanical and electric boundary conditions Eqs. (4.10) and (4.13) into Eq. (4.17) yields the following form for the energy imbalance:

$$\int_{\Omega_0} \dot{\Psi} dV \leq \int_{\Omega_0} (\mathbf{P} : \dot{\mathbf{F}} + \mathbf{E}_R \cdot \dot{\mathbf{D}}_R) \, dV + \int_{\Omega_0} (\text{Div}\mathbf{P} + \mathbf{B}) \cdot \dot{\mathbf{V}} dV + \int_{\Omega_0} \varphi \, (\text{-Div}\dot{\mathbf{D}}_R + \dot{\hat{q}}) \, dV. \tag{4.18}$$

It should be noticed in the above derivative, we also employ Eq. (4.2) and $\dot{\mathbf{F}} = \text{Grad}\dot{\chi}$.

Combining the mechanical equilibrium equation (Eq. 4.9) and Gauss's law (Eq. 4.12), this equation can be simplified to the following form at each local position as follows:

$$\dot{\Psi} - \mathbf{P} : \dot{\mathbf{F}} - \mathbf{E}_R \cdot \dot{\mathbf{D}}_R \leq 0. \tag{4.19}$$

The symmetric second Piola–Kirchhoff stress is related to **P** as

$$\mathbf{S} = \mathbf{F}^{-1}\mathbf{P} = J\mathbf{F}^{-1}\sigma\mathbf{F}^{-\text{T}}. \tag{4.20}$$

To represent the mechanical energy power, we can also use another thermodynamic pair **S** and **C** as

$$\mathbf{P} : \dot{\mathbf{F}} = \frac{1}{2}\mathbf{S} : \dot{\mathbf{C}}. \tag{4.21}$$

Thus, the free energy imbalance can also be expressed in another form as

$$\dot{\Psi} - \frac{1}{2}\mathbf{S} : \dot{\mathbf{C}} - \mathbf{E}_R \cdot \dot{\mathbf{D}}_R \leq 0. \tag{4.22}$$

The free energy is expressed as a function of right Cauchy–Green tensor **C**, nominal electric displacement $\mathbf{D}_R$, and internal variables $\mathbf{A}_i$ for viscoelastic relaxation effects as

$$\Psi = \Psi(\mathbf{C}, \mathbf{D}_R, \mathbf{A}_i). \tag{4.23}$$

The rate of free energy can then be written as

$$\dot{\Psi} = \frac{\partial \Psi}{\partial \mathbf{C}} : \dot{\mathbf{C}} + \frac{\partial \Psi}{\partial \mathbf{D}_R} : \dot{\mathbf{D}}_R + \sum \frac{\partial \Psi}{\partial \mathbf{A}_i} : \dot{\mathbf{A}}_i. \tag{4.24}$$

Substituting Eq. (4.24) into Eq. (4.22) gives

$$\left(2\frac{\partial \Psi}{\partial \mathbf{C}} - \mathbf{S}\right) : \frac{1}{2}\dot{\mathbf{C}} + \left(\frac{\partial \Psi}{\partial \mathbf{D}_R} - \mathbf{E}_R\right) \cdot \dot{\mathbf{D}}_R + \sum \frac{\partial \Psi}{\partial \mathbf{A}_i} : \dot{\mathbf{A}}_i \leq 0. \tag{4.25}$$

This inequality must hold for any values of **C** and $\mathbf{D}_R$, which requires the following relations to hold:

$$\mathbf{S} = 2\frac{\Phi(\vec{\mathbf{C}}, \vec{\mathbf{D}}_R, \vec{\mathbf{A}}_i)}{\partial \mathbf{C}}, \tag{4.26a}$$

$$\mathbf{E}_R = \frac{\Phi(\vec{\mathbf{C}}, \vec{\mathbf{D}}_R, \vec{\mathbf{A}}_i)}{\partial \mathbf{D}_R}. \tag{4.26b}$$

Thus, the second Piola–Kirchhoff stress and the nominal electric field can be obtained once the free energy is determined.

The remaining part of the energy imbalance can be further written as

$$\sum \frac{\partial \Psi}{\partial \mathbf{A}_i} : \dot{\mathbf{A}}_i \leq 0. \tag{4.27}$$

4.2.1.6 Free Energy Density of the Electric Part

The free energy of DEs is consisted of an equilibrium mechanical part $\Psi^{\text{eq}}_{\text{mech}}$, a nonequilibrium mechanical part $\Psi^{\text{neq}}_{\text{mech}}$, and an electric part Ψ_{ele}. Thus, the total free energy density can be represented as

$$\Psi = \Psi^{\text{eq}}_{\text{mech}} + \Psi^{\text{neq}}_{\text{mech}} + \Psi_{\text{ele}}. \tag{4.28}$$

To model the mechanical behaviors of elastomers, various hyperelastic and viscoelastic models have been introduced as discussed in Chapter 2. These models can be directly employed for the mechanical parts. Thus, it is a central task to obtain the corresponding energy form for the electric part.

An ideal dielectric model assumes that the permittivity is a constant, independent of deformation. Thus, the true electric field is related to the true electric displacement as

$$\mathbf{E}_{\text{c}} = \frac{\mathbf{D}}{\epsilon}, \tag{4.29}$$

where ϵ is the dielectric permittivity.

We have already obtained relations for electric field and electric displacement between reference and current configurations as shown in Eqs. (4.8) and (4.16). Substituting these equations into Eq. (4.29) leads to

$$\mathbf{E}_{\text{R}} = \frac{1}{\epsilon} J^{-1} \mathbf{C} \mathbf{D}_{\text{R}}. \tag{4.30}$$

Considering Eq. (4.26b), the free energy density of the electric part has the following form:

$$\Psi_{\text{ele}} = \frac{1}{2\epsilon} J^{-1} \mathbf{D}_{\text{R}} \mathbf{C} \mathbf{D}_{\text{R}}. \tag{4.31}$$

4.2.2 An Electro-hyperelastic Model

For this part, we first present an electro-hyperelastic model that does not consider the dissipation processes. Here, we employ the neo-Hookean model for the mechanical free energy, which is represented as

$$\Psi^{\text{eq}}_{\text{mech}} = \frac{\mu}{2}(I_1 - 3) - p(J - 1), \tag{4.32}$$

where I_1 is the first invariant of right Cauchy–Green tensor $\mathbf{C}$, $J = \det(\mathbf{F})$ and p is the hydrostatic pressure term to reinforce incompressibility.

Thus, the total free energy density can be represented as follows:

$$\Psi = \frac{\mu}{2}(I_1 - 3) + \frac{1}{2\epsilon} J^{-1} \mathbf{D}_{\text{R}} \mathbf{C} \mathbf{D}_{\text{R}} - p(J - 1). \tag{4.33}$$

Combining Eqs. (4.26a) and (4.33) leads to the following explicit form for the second Piola–Kirchhoff stress as follows:

$$\mathbf{S} = \mu\mathbf{I} + \frac{1}{\epsilon}\left[\mathbf{D}_{\mathrm{R}} \otimes \mathbf{D}_{\mathrm{R}} - \frac{1}{2}\mathbf{C}^{-1}\left(\mathbf{D}_{\mathrm{R}}\mathbf{C}\mathbf{D}_{\mathrm{R}}\right)\right] - p\mathbf{C}^{-1}. \tag{4.34}$$

The Cauchy stress tensor can be obtained as

$$\boldsymbol{\sigma} = \mu\mathbf{b} + \frac{1}{\epsilon}\left[\mathbf{F}\mathbf{D}_{\mathrm{R}} \otimes \mathbf{F}\mathbf{D}_{\mathrm{R}} - \frac{1}{2}\mathbf{F}\left(\mathbf{D}_{\mathrm{R}}\mathbf{C}\mathbf{D}_{\mathrm{R}}\right)\mathbf{F}^{-1}\right] - p\mathbf{I}. \tag{4.35}$$

Using Eqs. (4.8), (4.16), and (4.30), the Cauchy stress can also be expressed by spatial quantities as

$$\boldsymbol{\sigma} = \mu\mathbf{b} + \epsilon\left[\mathbf{E}_{\mathrm{c}} \otimes \mathbf{E}_{\mathrm{c}} - \frac{1}{2}\left(\mathbf{E}_{\mathrm{c}} \cdot \mathbf{E}_{\mathrm{c}}\right)\mathbf{I}\right] - p\mathbf{I}. \tag{4.36}$$

Consider a freestanding block as shown in Figure 4.1. An electric field along the thickness direction $\mathbf{e}_3$ with a true strength value of E is applied on it. Then, the three principal values σ_i are given as

$$\begin{aligned} \sigma_1 &= \mu\lambda_1^2 - \frac{\epsilon}{2}E^2 - p, \\ \sigma_2 &= \mu\lambda_2^2 - \frac{\epsilon}{2}E^2 - p, \\ \sigma_3 &= \mu\lambda_3^2 + \frac{\epsilon}{2}E^2 - p. \end{aligned} \tag{4.37}$$

Using the free traction condition in direction 3 to eliminate the pressure term, we can get that

$$\begin{aligned} \sigma_1 &= \mu\left(\frac{1}{\lambda_3} - \lambda_3^2\right) - \epsilon E^2, \\ \sigma_2 &= \mu\left(\frac{1}{\lambda_3} - \lambda_3^2\right) - \epsilon E^2, \\ \sigma_3 &= 0. \end{aligned} \tag{4.38}$$

In this equation, we have also assumed that the deformation in directions 1 and 2 are the same.

Using the above procedure, we can derive the explicit relations for DEs upon different loading conditions. The physical quantities in both reference and current configurations can also be determined.

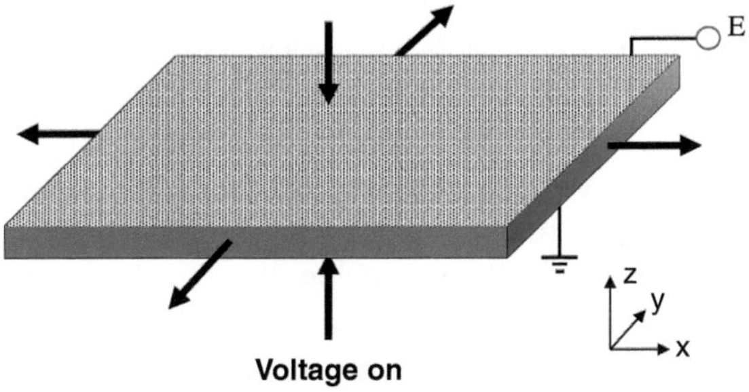

Figure 4.1 A freestanding DE block under an electric field in the z-direction.

4.2.3 An Electro-hyper-viscoelastic Model

For the most used acrylic elastomers, such as VHB elastomers, viscoelasticity is also an essential factor that affects performance. Thus, a visco-hyperelastic-electro constitutive model should be used to accurately predict the performance of these materials. Here we extend the above electro-hyperelastic model through incorporating the dissipation mechanism proposed by Linder et al. [35].

The nonequilibrium free energy is represented as

$$\Psi_{\text{mech}}^{\text{neq}} = \sum_i \frac{1}{2}\mu_i^{\text{neq}} \left(\mathbf{A}_i : \mathbf{C} - 3 - \ln\det\mathbf{A}_i\right), \tag{4.39}$$

where μ_i^{neq} is the nonequilibrium shear modulus for the ith relaxation process.

Following the discussion in Linder et al. [35], we adopt the following evolution equation for $\mathbf{A}_i$:

$$\dot{\mathbf{A}}_i = \frac{\mathbf{C}^{-1} - \mathbf{A}_i}{\tau_i}, \tag{4.40}$$

where τ_i is the relaxation time for the ith relaxation process. The above equation satisfies the entropy inequality (Eq. (4.27)) as shown in Linder et al. [35] and Chapter 2.

We still use the neo-Hookean model for the hyperelastic field and ideal dielectric model for the electric field. The total free energy is then given as

$$\Psi = \frac{\mu}{2}(I_1 - 3) + \sum_i \frac{1}{2}\mu_i^{\text{neq}}\left(\mathbf{A}_i : \mathbf{C} - 3 - \ln\det\mathbf{A}_i\right) + \frac{1}{2\epsilon}J^{-1}\mathbf{D}_{\text{R}}\mathbf{C}\mathbf{D}_{\text{R}} - p(J-1). \tag{4.41}$$

Similarly, we can have the explicit form of nominal and true stress as follows:

$$\mathbf{S} = \mu\mathbf{I} + \sum_i \mu_i^{\text{neq}}\mathbf{A}_i + \frac{1}{\epsilon}\left[\mathbf{D}_{\text{R}} \otimes \mathbf{D}_{\text{R}} - \frac{1}{2}\mathbf{C}^{-1}\left(\mathbf{D}_{\text{R}}\mathbf{C}\mathbf{D}_{\text{R}}\right)\right] - p\mathbf{C}^{-1}, \tag{4.42a}$$

$$\sigma = \mu\mathbf{b} + \sum_i \mu_i^{\text{neq}}\mathbf{F}\mathbf{A}_i\mathbf{F}^{\text{T}} + \epsilon\left[\mathbf{E}_{\text{c}} \otimes \mathbf{E}_{\text{c}} - \frac{1}{2}\left(\mathbf{E}_{\text{c}} \cdot \mathbf{E}_{\text{c}}\right)\mathbf{I}\right] - p\mathbf{I}. \tag{4.42b}$$

4.2.4 Electromechanical Instability of Dielectric Elastomers

When a DE material is actuated under an electric field as shown in Figure 4.1, the actuated DE thins on the lateral direction. As the electric field further increases, the actuator becomes thinner, leading to a further increase in the electric field. When the electric field reaches a critical value, the actuator undergoes a dramatic, abrupt deformation known as electromechanical instability (EMI), typically accompanied by subsequent electric breakdown. Various analysis and methods have been proposed to eliminate the EMI; an effective way is to apply a pre-stretch on the DE actuator before applying an electric field [32–34]. The pre-stretch places the DEs in the strain-stiffening region, thereby enhancing their resistance against further thickness reduction. The actuator can then sustain a large deformation under a relatively large electric field strength. In this part, we perform a theoretical analysis from the perspective of energy to determine the deformation state of the DE actuator. We consider a rectangular DE block with the dimension of $L_1L_2L_3$, subjected to an electric voltage (potential) φ and two mechanical forces G_1 and G_2 as shown in Figure 4.2.

The DE material is assumed to be an incompressible material with $\lambda_1\lambda_2\lambda_3 = 1$. The block has a surface electric charge of ω. The deformed lengths are denoted as $\lambda_i L_i$. The nominal electric field and the nominal electric displacement are

$$E_{R3} = \frac{\varphi}{L_3}, \tag{4.43a}$$

$$D_{R3} = \frac{\omega}{L_1L_2}. \tag{4.43b}$$

The true electric field and displacement are

$$E_3 = \frac{\varphi}{\lambda_3 L_3}, \tag{4.44a}$$

$$D_3 = \frac{\omega}{\lambda_1 L_1 \lambda_2 L_2}. \tag{4.44b}$$

The nominal stress in the principal directions is represented as

$$P_1 = \frac{G_1}{L_2L_3}, \tag{4.45a}$$

$$P_2 = \frac{G_2}{L_1L_3}, \tag{4.45b}$$

$$P_3 = 0. \tag{4.45c}$$

The hyperelastic free energy density can be a function of stretches. Considering the incompressibility condition, the hyperelastic free energy can be further assumed to be dependent on the first two principal stretches, λ_1 and λ_2. Therefore, the total free energy

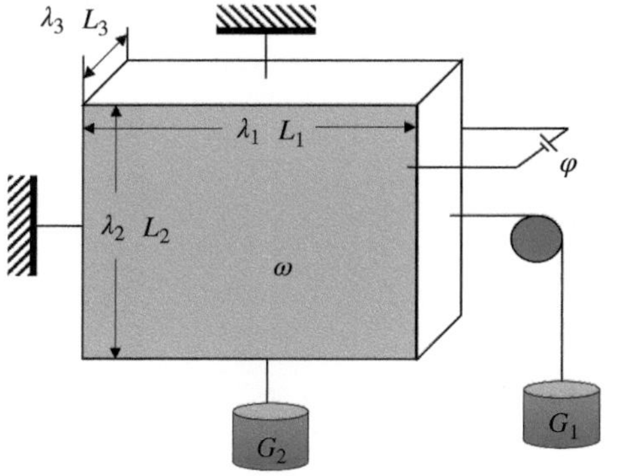

Figure 4.2 A DE block coated with two compliant electrodes and subjected to an electric voltage of φ through the thickness and two mechanical loads G_1 and G_2.

density Ψ can be represented as $\Psi(\lambda_1, \lambda_2, D_3)$. Considering the extra work done by the electric and mechanical loads, the free energy W of the whole system can be written as

$$W = L_1 L_2 L_3 \Psi(\lambda_1, \lambda_2, D_3) - G_1 \lambda_1 L_1 - G_2 \lambda_2 L_2 - \varphi \omega. \tag{4.46}$$

The second-order variation of the energy per unit reference volume is

$$\begin{aligned} \frac{\delta W}{L_1 L_2 L_3} &= \left(\frac{\partial \Psi}{\partial \lambda_1} - P_1\right)\delta\lambda_1 + \left(\frac{\partial \Psi}{\partial \lambda_2} - P_2\right)\delta\lambda_2 + \left(\frac{\partial \Psi}{\partial D_3} - E_R\right)\delta D_3 \\ &+ \frac{1}{2}\frac{\partial^2 \Psi}{\partial \lambda_1^2}\delta\lambda_1^2 + \frac{1}{2}\frac{\partial^2 \Psi}{\partial \lambda_2^2}\delta\lambda_2^2 + \frac{1}{2}\frac{\partial^2 \Psi}{\partial D_3^2}\delta D_3^2 \\ &+ \frac{\partial^2 \Psi}{\partial \lambda_1 \partial \lambda_2}\delta\lambda_1\delta\lambda_2 + \frac{\partial^2 \Psi}{\partial \lambda_1 \partial D_3}\delta\lambda_1\delta D_3 + \frac{\partial^2 \Psi}{\partial \lambda_2 \partial D_3}\delta\lambda_2\delta D_3. \end{aligned} \tag{4.47}$$

The equilibrium under the thermodynamic framework requires the minimum free energy density. If the system is under an equilibrium state, the first-order variations terms vanish, which requires that

$$P_1 = \frac{\partial \Psi}{\partial \lambda_1}, \quad P_2 = \frac{\partial \Psi}{\partial \lambda_2}, \quad E_R = \frac{\partial \Psi}{\partial D_3}. \tag{4.48}$$

To ensure the free energy is minimum, the sum of second-order variations must be positive for arbitrary $\delta\lambda_1$,$\delta\lambda_2$, and δD_3, which requires a positive definite of the Hessian matrix, defined as

$$\mathbf{H} = \begin{bmatrix} \dfrac{\partial^2 \Psi}{\partial \lambda_1^2} & \dfrac{\partial^2 \Psi}{\partial \lambda_1 \partial \lambda_2} & \dfrac{\partial^2 \Psi}{\partial \lambda_1 \partial D_3} \\ \dfrac{\partial^2 \Psi}{\partial \lambda_1 \partial \lambda_2} & \dfrac{\partial^2 \Psi}{\partial \lambda_2^2} & \dfrac{\partial^2 \Psi}{\partial \lambda_2 \partial D_3} \\ \dfrac{\partial^2 \Psi}{\partial \lambda_1 \partial D_3} & \dfrac{\partial^2 \Psi}{\partial \lambda_2 \partial D_3} & \dfrac{\partial^2 \Psi}{\partial D_3^2} \end{bmatrix}. \tag{4.49}$$

We use the neo-Hookean model and the ideal dielectric model for DE materials. The free energy function is given as

$$\Psi(\lambda_1, \lambda_2, D_3) = \frac{\mu}{2}\left(\lambda_1^2 + \lambda_2^2 + \frac{1}{\lambda_1^2 \lambda_2^2} - 3\right) + \frac{D_3}{2\epsilon}\frac{1}{\lambda_1^2 \lambda_2^2}. \tag{4.50}$$

The equilibrium equations described by stretches are

$$P_1 = \mu\left(\lambda_1 - \frac{1}{\lambda_1^3 \lambda_2^2}\right) - \frac{D_3^2}{\epsilon}\frac{1}{\lambda_1^3 \lambda_2^2}, \tag{4.51a}$$

$$P_2 = \mu\left(\lambda_2 - \frac{1}{\lambda_2^3 \lambda_1^2}\right) - \frac{D_3^2}{\epsilon}\frac{1}{\lambda_2^3 \lambda_1^2}, \tag{4.51b}$$

$$E_R = \frac{D_3}{\epsilon}\frac{1}{\lambda_1^2 \lambda_2^2}. \tag{4.51c}$$

The Hessian can be written in the following form:

$$\mathbf{H}=\begin{bmatrix} \mu\left(1+3\frac{1}{\lambda_1^4\lambda_2^2}\right)+\frac{3D_3^2}{\epsilon}\frac{1}{\lambda_1^4\lambda_2^2} & 2\mu\frac{1}{\lambda_1^3\lambda_2^3}+\frac{2D_3^2}{\epsilon}\frac{1}{\lambda_1^2\lambda_2^3} & -\frac{2D_3}{\epsilon}\frac{1}{\lambda_1^3\lambda_2^2} \\ 2\mu\frac{1}{\lambda_1^3\lambda_2^3}+\frac{2D_3^2}{\epsilon}\frac{1}{\lambda_1^2\lambda_2^3} & \mu\left(1+3\frac{1}{\lambda_2^4\lambda_1^2}\right)+\frac{3D_3^2}{\epsilon}\frac{1}{\lambda_2^4\lambda_1^2} & -\frac{2D_3}{\epsilon}\frac{1}{\lambda_2^3\lambda_1^2} \\ -\frac{2D_3}{\epsilon}\frac{1}{\lambda_1^3\lambda_2^2} & -\frac{2D_3}{\epsilon}\frac{1}{\lambda_2^3\lambda_1^2} & \frac{1}{\epsilon}\frac{1}{\lambda_1^2\lambda_2^2} \end{bmatrix}. \tag{4.52}$$

The instability can be determined by examining the Hessian. Consider a simple loading condition. The block is mechanical load free, and the deformation state is equal to biaxial deformation, which gives $\lambda_1=\lambda_2=\lambda$ and $G_1=G_2=0$. The constitutive relation of Eq. (4.51) gives

$$D_3=\sqrt{\epsilon\mu}\sqrt{\lambda^6-1}. \tag{4.53}$$

The true stress can be then represented as

$$\sigma=\mu\left(\lambda^2-\frac{1}{\lambda^4}\right)-\epsilon E^2. \tag{4.54}$$

With the freestanding condition $\sigma=0$, we can have the explicit equation for the applied voltage against stretch:

$$\hat{\varphi}=\lambda^{-2}\sqrt{\lambda^2-\frac{1}{\lambda^4}}, \tag{4.55}$$

where $\hat{\varphi}$ is nondimensionalized voltage with $\hat{\varphi}=\sqrt{\frac{\epsilon}{\mu}}\frac{\varphi}{L_3}$. Assume the thickness L_3 equals to 1 mm, and using the material parameters of VHB in Mathew et al. [29], we can calculate the determinant of Hessian det(**H**) for arbitrary deformation state, which is plotted in Figure 4.3a. The corresponding nondimensionalized voltage against the stretch is shown in Figure 4.3b.

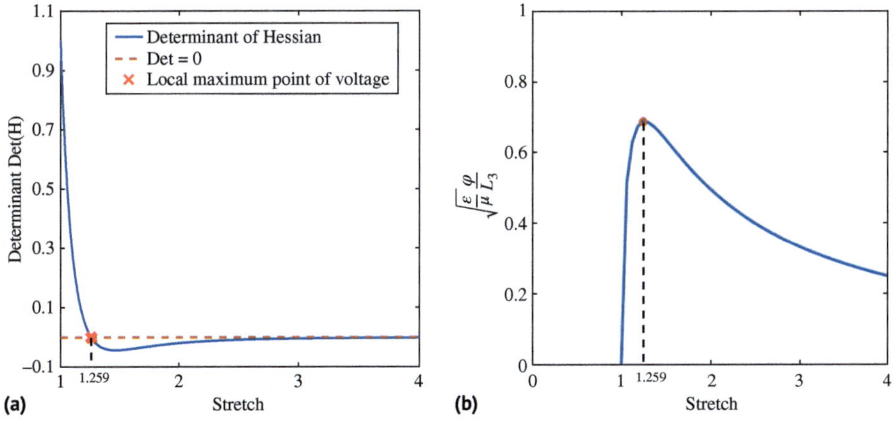

Figure 4.3 (a) The determinant value against stretch and (b) the applied voltage against stretch.

The local maximum point of voltage in Figure 4.3b is apparently the critical point, where the determinant of Hessian will turn from positive to negative when the deformation surpasses the local maximum point of voltage. For the neo-Hookean model, the critical stretch is calculated as $\lambda_c = 2^{1/3} \approx 1.26$. This critical stretch varies for different models. Generally, instability will happen when the stretch exceeds the critical value. It has been found that pre-stretch can effectively eliminate the EMI and induce the large deformation capacity. Recently, Shi et al. [36] reported a bimodal network elastomer system, which shows an apparent strain stiffening effect in the median strain regime. These elastomers do not exhibit the EMI phenomenon during electric loading. Thus, these elastomers can be used in actuators without needing the constraints from the rigid frame.

4.3 Finite Element Implementation

In this part, we show the general procedures to implement the continuum model for DEs into COMSOL Multiphysics. For analytical studies, it is relatively simple to use the incompressible assumption. For finite element simulation, it is more convenient to use the quasi-incompressible condition. Thus, we first extend the above model to the quasi-incompressible condition. We then derive the weak form for the finite element implementation.

4.3.1 Constitutive Relation

The deformation gradient is first decomposed into a volumetric part and a deviatoric part as

$$\bar{\mathbf{F}} = J^{-\frac{1}{3}}\mathbf{F}. \tag{4.56}$$

The deviatoric right Cauchy–Green tensor is defined as

$$\bar{\mathbf{C}} = J^{-\frac{2}{3}}\mathbf{C}, \quad \bar{\mathbf{b}} = J^{-\frac{2}{3}}\mathbf{b}. \tag{4.57}$$

The first invariant of $\bar{\mathbf{C}}$ is denoted as $\bar{I}_1$.

The total free energy density can then be represented as

$$\begin{aligned}\Psi = \frac{\mu}{2}(\bar{I}_1 - 3) + \sum_i \frac{1}{2}\mu_i^{\text{neq}}\left(\mathbf{A}_i : \bar{\mathbf{C}} - 3 - \ln\det\mathbf{A}_i\right) + \frac{\kappa}{2}(J-1)^2 \\ + \frac{1}{2\epsilon}J^{-1}\mathbf{D}_{\mathrm{R}}\bar{\mathbf{C}}\mathbf{D}_{\mathrm{R}},\end{aligned} \tag{4.58}$$

where κ is the bulk modulus.

By combining Eqs. (4.26) and (4.58), the stress and the electric displacement can then be calculated as

$$\begin{aligned}\mathbf{P} = \mu J^{-2/3}\left(\mathbf{F} - \frac{1}{3}I_1\mathbf{F}^{-\mathrm{T}}\right) + \sum_i \mu_i^{\text{neq}} J^{-2/3}\left[\mathbf{F}\mathbf{A}_i - \frac{1}{3}(\mathbf{A}_i : \mathbf{C})\,\mathbf{F}^{-\mathrm{T}}\right] \\ + \kappa J(J-1)\mathbf{F}^{-\mathrm{T}} + \frac{1}{\epsilon}J^{-1}\mathbf{F}\left[\mathbf{D}_{\mathrm{R}} \otimes \mathbf{D}_{\mathrm{R}} - \frac{1}{2}(\mathbf{D}_{\mathrm{R}}\mathbf{C}\mathbf{D}_{\mathrm{R}})\,\mathbf{C}^{-1}\right],\end{aligned} \tag{4.59a}$$

$$\mathbf{S}=\mu J^{-2/3}\left(\mathbf{I}-\frac{1}{3}I_1\mathbf{C}^{-1}\right)+\sum_i \mu_i^{\text{neq}}J^{-2/3}\left[\mathbf{A}_i-\frac{1}{3}(\mathbf{A}_i:\mathbf{C})\,\mathbf{C}^{-1}\right] \\ +\kappa J(J-1)\mathbf{C}^{-1}+\frac{1}{\epsilon}J^{-1}\left[\mathbf{D}_\mathrm{R}\otimes\mathbf{D}_\mathrm{R}-\frac{1}{2}(\mathbf{D}_\mathrm{R}\mathbf{C}\mathbf{D}_\mathrm{R})\,\mathbf{C}^{-1}\right], \tag{4.59b}$$

$$\sigma=\frac{1}{J}\mu\left(\bar{\mathbf{b}}-\frac{1}{3}\bar{I}_1\mathbf{I}\right)+\frac{1}{J}\sum_i \mu_i^{\text{neq}}\left[\bar{\mathbf{F}}\mathbf{A}_i\bar{\mathbf{F}}^\mathrm{T}-\frac{1}{3}(\mathbf{A}_i:\bar{\mathbf{C}})\,\mathbf{I}\right] \\ +\kappa(J-1)\mathbf{I}+\epsilon\left[\mathbf{E}_\mathrm{c}\otimes\mathbf{E}_\mathrm{c}-\frac{1}{2}(\mathbf{E}_\mathrm{c}\cdot\mathbf{E}_\mathrm{c})\,\mathbf{I}\right], \tag{4.59c}$$

$$\mathbf{E}_\mathrm{R}=\frac{1}{\epsilon}J^{-1}\mathbf{C}\mathbf{D}_\mathrm{R}. \tag{4.59d}$$

The corresponding evolution equation for $\mathbf{A}_i$ can be rewritten as

$$\dot{\mathbf{A}}_i=\frac{\bar{\mathbf{C}}^{-1}-\mathbf{A}_i}{\tau_i}. \tag{4.60}$$

4.3.2 Weak Forms

For the mechanical part, when the inertial force and body force are neglected, the balance of momentum can be written as

$$\mathrm{Div}\mathbf{P}=0. \tag{4.61}$$

The boundary condition is specified as Eq. (4.10), the weighting function is denoted as $\mathbf{w}_1$. Thus, the weak form can be written as

$$\int_{\Omega_0}\mathbf{P}:\mathrm{Grad}\,\mathbf{w}_1\,\mathrm{dV}-\int_{\partial\Omega_0}\mathbf{T}\cdot\mathbf{w}_1\,\mathrm{dS}=0. \tag{4.62}$$

The above weak form can be solved in the Solid Mechanics module.

For the electric field, Gauss's law Eq. (4.12) is the strong form. The boundary condition is specified as shown in Eq. (4.13). The weighting function is chosen as $\mathbf{w}_2$. We can then obtain the corresponding weak form for the electric field:

$$\int_{\Omega_0}\mathbf{D}_\mathrm{R}\mathbf{w}_2\,\mathrm{dV}+\int_{\partial\Omega_0}\hat{\omega}\mathbf{w}_2\,\mathrm{dS}=0. \tag{4.63}$$

Combining Eq. (4.2) with the above equation, the weak form for the electric field is

$$\int_{\Omega_0}\mathbf{F}^{-1}\mathrm{Grad}\,\varphi\cdot\mathbf{F}^{-1}\mathrm{Grad}\,\mathbf{w}_2\,\mathrm{dV}=0. \tag{4.64}$$

The above weak form can be solved by Weak Form PDEs interface in COMSOL.

For the viscoelastic part, the evolution equation of the internal variable in Eq. (4.40) can be solved using the coefficient form PDE interfaces in COMSOL. Since the internal variable is symmetric, only six components are independent. Moreover, Eq. (4.40) shows that each component is independent of others, which significantly reduces the computational cost.

4.3.2.1 Simulation Examples

In this part, we show the general performance of finite element simulations. The model parameters are set as below: the shear modulus for neo-Hookean part μ equals to 0.4 MPa and the volumetric modulus κ equals to 40 MPa ensuring the quasi-incompressible assumption, the nonequilibrium shear modulus for one relaxation process μ^{neq} equals to 15 MPa, relaxation time $\tau = 0.002$ s, the dielectric permittivity $\epsilon = 5 \times 10^{-11}\text{Fm}^{-1}$. Before simulating a relatively complex demo, we first verify whether the simulated results agree with the theory.

Here we first simulate a static loading, where the viscoelastic behavior can be ignored. In this situation, the second term in Eq. (4.58) describing viscoelasticity vanishes. We consider a mechanical-load-free square block with the dimension of 1 mm × 1 mm × 1 mm shown in Figure 4.4, similar to that in Figure 4.1. The theoretical analysis is under the incompressible assumption for convenience. Thus, the electrically stimulated Cauchy stress is the same as in Eq. (4.37). When we use the traction-free boundary condition in direction 1 and direction 2, the total stress in direction 3 becomes

$$\sigma_3 = \mu\left(\lambda_3^2 - \frac{1}{\lambda_3}\right) + \epsilon E^2, \tag{4.65}$$

where E is the true electric field strength, represented as

$$E = \frac{\varphi}{\lambda_3 H}, \tag{4.66}$$

where H is the thickness of the block, with $H = 1$.

When the block is in equilibrium, the total stress equals to zero, using Eq. (4.66), we can obtain

$$\lambda_3^4 - \lambda_3 + \frac{\epsilon\varphi^2}{\mu H^2} = 0. \tag{4.67}$$

By solving above equation, we can calculate the stretch λ_3 under a certain voltage, theoretically. Figure 4.5a shows the simulated displacement in the z-direction of a block under

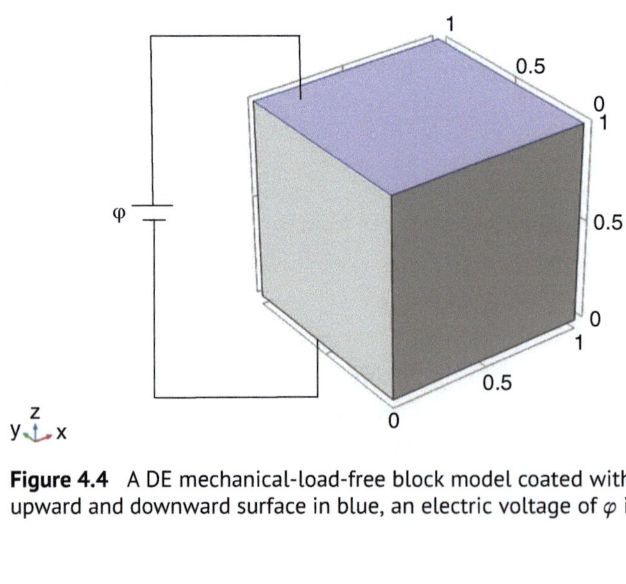

Figure 4.4 A DE mechanical-load-free block model coated with two compliant electrodes on the upward and downward surface in blue, an electric voltage of φ is applied.

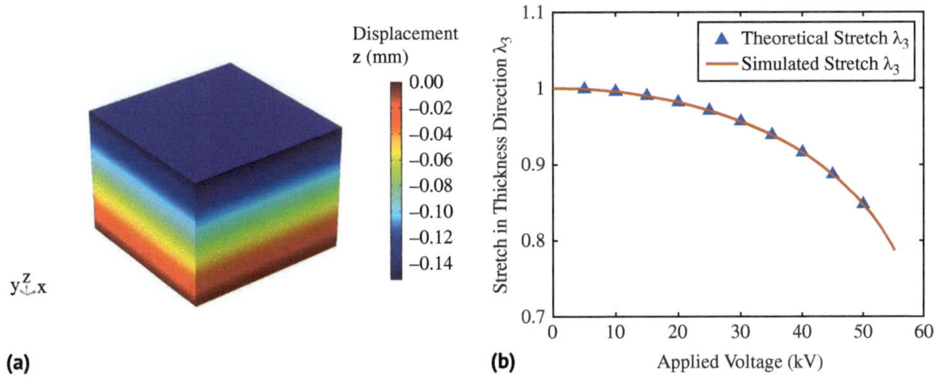

Figure 4.5 (a) Contour plot for displacement z of the block and (b) comparison between simulated and theoretical stretch λ_3.

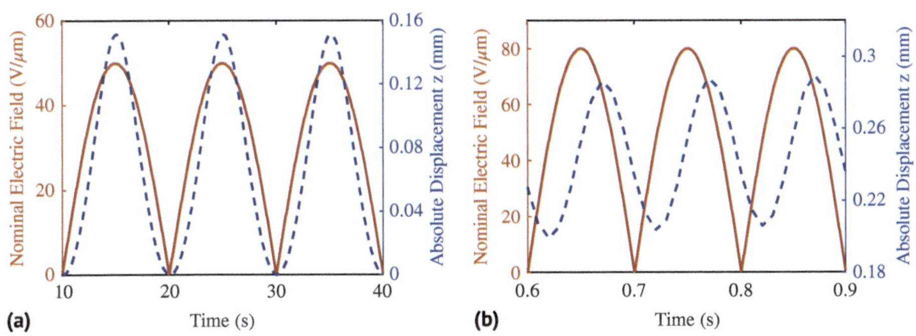

Figure 4.6 The electroactivated response of the block upon a sinusoidal voltage at a frequency of (a) 0.1 Hz and (b) comparison between simulated and theoretical stretch λ_3 of the block.

an electric field strength of 50 V/μm. Changing the applied voltage from 5 to 50 kV, the simulated and theoretical calculated stretch λ_3 is shown in Figure 4.5b. As shown, simulated results agree with the theoretical results. It has been demonstrated that when the volumetric modulus κ is larger enough (100 times of the shear modulus here), the quasi-incompressible hyperelastic model yields almost identical results to the incompressible model.

Further, we simulate the frequency-dependent electroactivated response upon dynamic voltage loadings. The absolute value of the displacement in the z-direction at a reference point (1,1,1) is selected to represent the deformation, as displayed in Figure 4.6. In the low-frequency situation, 0.1 Hz, the maximum applied voltage is 50 V/μm, while in the high frequency, 10 Hz, the maximum voltage is 80 V/μm. When the frequency is high, the response has a lag phenomenon with the load; therefore, the deformation is relatively smaller. To make the deformation more obvious, we set the maximum voltage to a larger value for higher frequencies. As shown in Figure 4.6a, the deformation is in phase with the applied voltage at low frequencies. When the frequency is 10 Hz in Figure 4.6b, the deformation lags with a phase shift of around 0.2 π.

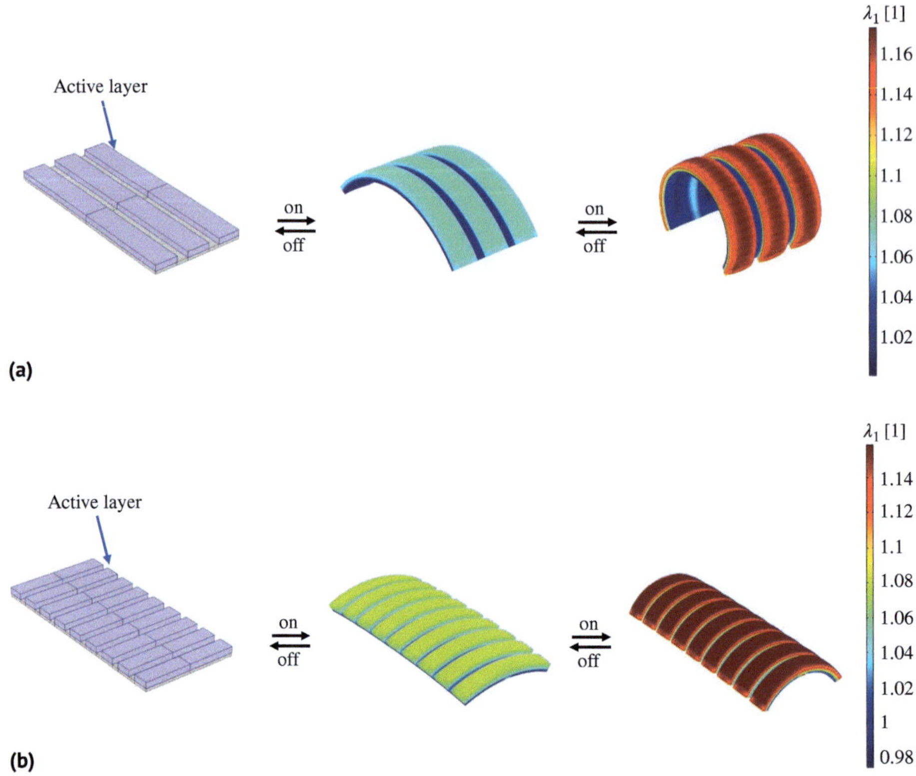

Figure 4.7 The electroactivated bending modes decided by the direction of length direction of active beams.

Next, we design a $3 \times 6.5 \times 0.1$ mm plate, attaching a series of linearly distributed active beams. The width and thickness of the beams are 0.75 and 0.2 mm, with a spacing of 0.375 mm, as shown in blue in Figure 4.7. The neo-Hookean model with the same parameters used above is used for DEs. The passive plate is also characterized by the neo-Hookean model with the shear modulus equal to 0.4 MPa. When the voltage is applied, the active beams expand in both length and width directions. As a result, different bending modes can be achieved by varying the distributions of the beams.

4.4 Conclusion

In this part, we presented a general process to build an electromechanical constitutive model for DE materials under the thermodynamic framework. Viscoelasticity is also considered in the model. Based on this model, we further apply a theoretical analysis method by calculating the determinant of Hessian, which can be employed to determine the EMI, a common failure mode in the applications of DE materials. A critical stretch can be identified as a threshold for EMI. At last, we implemented the electromechanical constitutive model for finite element analysis. A DE-based soft gripper is then simulated. The chapter provides a feasible framework for simulating the electroactivated response in elastomers.

References

1 Yufeng Chen, Huichan Zhao, Jie Mao, Pakpong Chirarattananon, E Farrell Helbling, Nak-seung Patrick Hyun, David R Clarke, and Robert J Wood. Controlled flight of a microrobot powered by soft artificial muscles. *Nature*, 575(7782): 324–329, 2019.

2 Chongjing Cao, Xing Gao, and Andrew T Conn. A magnetically coupled dielectric elastomer pump for soft robotics. *Advanced Materials Technologies*, 4(8): 1900128, 2019.

3 Guo-Ying Gu, Jian Zhu, Li-Min Zhu, and Xiangyang Zhu. A survey on dielectric elastomer actuators for soft robots. *Bioinspiration & Biomimetics*, 12(1): 011003, 2017.

4 Tongqing Lu, Cheng Ma, and Tiejun Wang. Mechanics of dielectric elastomer structures: a review. *Extreme Mechanics Letters*, 38: 100752, 2020.

5 Yuxuan Guo, Qicong Qin, Ziqing Han, Roshan Plamthottam, Mason Possinger, and Qibing Pei. Dielectric elastomer artificial muscle materials advancement and soft robotic applications. *SmartMat*, 4(4): e1203, 2023.

6 Xinghai Pan, Wei Pu, Yanling Liu, Yuhang Xiao, Junhong Pu, Ye Shi, Hui Wu, and Haolun Wang. Self-perceptional soft robotics by a dielectric elastomer. *ACS Applied Materials & Interfaces*, 16(20): 26797–26807, 2024.

7 Chao Tang, Boyuan Du, Songwen Jiang, Zhong Wang, Xin-Jun Liu, and Huichan Zhao. A review on high-frequency dielectric elastomer actuators: materials, dynamics, and applications. *Advanced Intelligent Systems*, 6(2): 2300047, 2024.

8 Wenwen Feng, Lin Sun, Zhekai Jin, Lili Chen, Yuncong Liu, Hao Xu, and Chao Wang. A large-strain and ultrahigh energy density dielectric elastomer for fast moving soft robot. *Nature Communications*, 15(1): 4222, 2024.

9 Krishna Veer Singh Gurjar, Anup Sankar Sadangi, Ajeet Kumar, Dilshad Ahmad, Karali Patra, Ieuan Collins, Mokarram Hossain, Rafic M Ajaj, and Yahya Zweiri. Dielectric elastomer generators: recent advances in materials, electronic circuits, and prototype developments. *Advanced Energy and Sustainability Research*, 6(1): 2400221, 2025.

10 Samuel Shian, Katia Bertoldi, and David R Clarke. Dielectric elastomer based "grippers" for soft robotics. *Advanced Materials*, 27(43): 6814–6819, 2015.

11 Yanze Liu, Shangzhi Yue, Zhongyuan Tian, Zijuan Zhu, Yongji Li, Xiangyu Chen, Zhong Lin Wang, Zhong-Zhen Yu, and Dan Yang. Self-powered and self-healable extraocular-muscle-like actuator based on dielectric elastomer actuator and triboelectric nanogenerator. *Advanced Materials*, 36(7): 2309893, 2024.

12 Siegfried Bauer, Simona Bauer-Gogonea, Ingrid Graz, Martin Kaltenbrunner, Christoph Keplinger, and Reinhard Schwödiauer. 25th anniversary article: a soft future: from robots and sensor skin to energy harvesters. *Advanced Materials*, 26(1): 149–162, 2014.

13 Yaguang Guo, Liwu Liu, Yanju Liu, and Jinsong Leng. Review of dielectric elastomer actuators and their applications in soft robots. *Advanced Intelligent Systems*, 3(10): 2000282, 2021.

14 Paul Brochu and Qibing Pei. Advances in dielectric elastomers for actuators and artificial muscles. *Macromolecular Rapid Communications*, 31(1): 10–36, 2010.

15 Le An, Fangfang Wang, Sibo Cheng, Tongqing Lu, and T J Wang. Experimental investigation of the electromechanical phase transition in a dielectric elastomer tube. *Smart Materials and Structures*, 24(3): 035006, 2015.

16 Godaba, Hareesh Foo, Choon Chiang Zhang, Zhi Qian Khoo, and Boo Cheong Zhu. Jian Giant voltage-induced deformation of a dielectric elastomer under a constant pressure. *Applied Physics Letters*, 105(11): 112901, 2014.

17 Robert M McMeeking and Chad M Landis. Electrostatic forces and stored energy for deformable dielectric materials. *Journal of Applied Mechanics*, 72(4): 581–590, 2005.

18 Zhigang Suo, Xuanhe Zhao, and William H Greene. A nonlinear field theory of deformable dielectrics. *Journal of the Mechanics and Physics of Solids*, 56(2): 467–486, 2008.

19 Xuanhe Zhao and Zhigang Suo. Theory of dielectric elastomers capable of giant deformation of actuation. *Physical review letters*, 104(17): 178302, 2010.

20 Choon Chiang Foo, Shengqiang Cai, Soo Jin Adrian Koh, Siegfried Bauer, and Zhigang Suo. Model of dissipative dielectric elastomers. *Journal of Applied Physics*, 111(3): 034102, 2012.

21 Harold S Park and Thao D Nguyen. Viscoelastic effects on electromechanical instabilities in dielectric elastomers. *Soft Matter*, 9(4): 1031–1042, 2013.

22 Markus Mehnert, Jessica Faber, Mokarram Hossain, Shawn A Chester, and Paul Steinmann. Experimental and numerical investigations of the electro-mechanical response of particle filled elastomers—part II: continuum modeling approach. *European Journal of Mechanics-A/Solids*, 96: 104661, 2022.

23 Lingling Chen, Xu Yang, Chuo Zhao, and Shengyou Yang. Electro-mechanical surface wrinkling of a finite block of dielectric elastomers accounting for the pre-stretch and the aspect ratio. *International Journal of Engineering Science*, 198: 104041, 2024.

24 Lu Dai, Junwei Xu, and Rui Xiao. Modeling the stimulus-responsive behaviors of fiber-reinforced soft materials. *International Journal of Applied Mechanics*, 16(6): 2450041, 2024.

25 Huang, Weiyang Chen, Kaijuan Ma, and Pengyu Kang. Guozheng A visco-hyperelastic constitutive model for temperature-dependent cyclic deformation of dielectric elastomer. *Acta Mechanica Solida Sinica*, 37: 1–14, 2024.

26 Rui Xiao, Zike Chen, Ye Shi, Lin Zhan, Shaoxing Qu, and Paul Steinmann. A continuum model for novel electromechanical-instability-free dielectric elastomers. *Journal of the Mechanics and Physics of Solids*, 196: 105994, 2025.

27 David L Henann, Shawn A Chester, and Katia Bertoldi. Modeling of dielectric elastomers: design of actuators and energy harvesting devices. *Journal of the Mechanics and Physics of Solids*, 61(10): 2047–2066, 2013.

28 Markus Mehnert, Mokarram Hossain, and Paul Steinmann. A complete thermo-electro-viscoelastic characterization of dielectric elastomers, part II: continuum modeling approach. *Journal of the Mechanics and Physics of Solids*, 157: 104625, 2021.

29 Anup Teejo Mathew, Tran Vy Khanh Vo, and Soo Jin Adrian Koh. A molecular perspective to analytical modeling that reveals new instabilities in dielectric elastomer transducers. *Journal of the Mechanics and Physics of Solids*, 132: 103703, 2019.

30 Sankalp Gour and Deepak Kumar. Thermo-electrostrictive unequal-biaxial deformation instability of electro-active polymeric composites. *International Journal of Non-Linear Mechanics*, 156: 104481, 2023.

31 Shengyou Yang and Pradeep Sharma. A tutorial on the stability and bifurcation analysis of the electromechanical behaviour of soft materials. *Applied Mechanics Reviews*, 75(4): 044801, 2023.

32 Yipin Su, Weiqiu Chen, and Michel Destrade. Tuning the pull-in instability of soft dielectric elastomers through loading protocols. *International Journal of Non-Linear Mechanics*, 113: 62–66, 2019.

33 Bo Li and Zijie Zhao. Electromechanical deformation of dielectric elastomer in two types of pre-stretch. *Europhysics Letters*, 106(6): 67009, 2014.
34 Liang Jiang, Yanfen Zhou, Shaojuan Chen, Jianwei Ma, Anthony Betts, and Stephen Jerrams. Electromechanical instability in silicone-and acrylate-based dielectric elastomers. *Journal of Applied Polymer Science*, 135(9): 45733, 2018.
35 Christian Linder, Mykola Tkachuk, and Christian Miehe. A micromechanically motivated diffusion-based transient network model and its incorporation into finite rubber viscoelasticity. *Journal of the Mechanics and Physics of Solids*, 59(10): 2134–2156, 2011.
36 Ye Shi, Erin Askounis, Roshan Plamthottam, Tom Libby, Zihang Peng, Kareem Youssef, Junhong Pu, Ron Pelrine, and Qibing Pei. A processable, high-performance dielectric elastomer and multilayering process. *Science*, 377(6602): 228–232, 2022.

5

A Magnetomechanical Coupled Model for Magnetoactive Soft Materials

Magnetoactive soft materials are smart composites that integrate magnetic particles within a soft matrix, endowing them with stimulus-responsive behaviors upon the magnetic field. In this chapter, we present a magnetomechanical coupled model for magnetorheological elastomers reinforced with hard magnetic particles, which has received considerable interest due to the feasible fabrication process via 3D printing. We also compare the performance of two models with different forms of magnetic energy to predict the responsive behaviors of magnetoactive soft materials. Finally, we incorporate the glass transition mechanism into the model, which can be used to describe the behaviors of magnetic activated shape-memory polymers, which can be programmed above the glass transition temperature upon the magnetic field and fixed in the temporary shape due to glass transition upon cooling.

5.1 Introduction

Magnetoactive soft materials, such as magnetorheological elastomers (MREs) and ferrogels, are composite structures typically consisting of magnetic particles (iron or iron oxides) embedded within polymeric matrices. The pioneering work of Rigbi and Jilken [1] fabricated magnetoactive particle-filled MREs by embedding soft iron particles within a natural rubber matrix, and subsequently examined the mechanical properties of the composite under the applied magnetic field. In general, magnetic particles embedded into soft materials are classified based on their coercivity as either soft magnetic or hard magnetic, each offering distinct functional advantages in response to magnetic fields.

Soft magnetic particles (low-coercivity ferromagnetic particles), such as iron and iron oxides, possess low coercivity, enabling them to develop strong magnetization when exposed to an external magnetic field. However, due to their low coercivity (represented by low $\mathbf{H}^c$ as shown in Figure 5.1), they retain minimal residual magnetization once the magnetic field is removed. When incorporated into a compliant polymer matrix, the resulting material is defined as soft-magnetic-particle-filled magnetorheological elastomers (s-MREs), which are primarily constrained to simple elongation or compression under an external magnetic field. In these composites, magnetic interactions between particles and the surrounding polymer matrix allow for controlled tuning of mechanical properties, such as modulus and stiffness. A range of fabrication techniques and experimental probes for

Multi-field Modeling of Soft Active Materials: Properties and Design, First Edition. Rui Xiao.

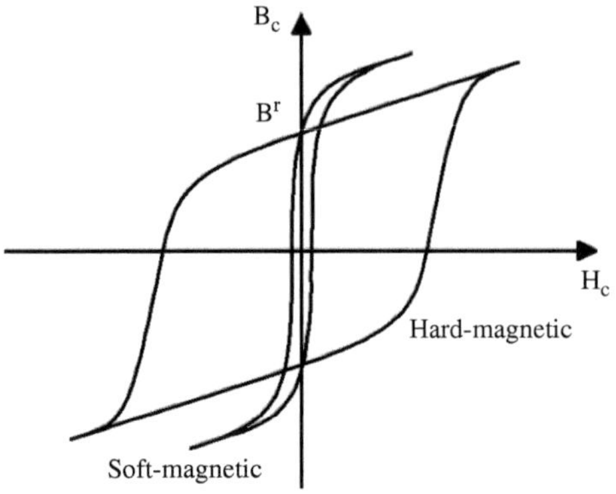

Figure 5.1 Magnetic hysteresis loops, and $\mathbf{B}_c - \mathbf{H}_c$ curve of soft magnetic and hard magnetic particles.

s-MREs have been reported [2–5]. For example, Chen et al. [5] fabricated a magnetic hydrogel octopus by using AAm-Carbomer ink mixed with ferromagnetic particles (Fe_3O_4), which can be actuated by a programmable magnetic field.

In contrast, hard magnetic particles (high-coercivity ferromagnetic particles), such as neodymium-iron-boron (NdFeB), retain high residual magnetic flux density (represented by high $\mathbf{B}^r$ in Figure 5.1) even in the absence of external magnetic fields once magnetically saturated. Hard magnetic iron-particle-filled magnetorheological elastomers (h-MREs) are well suited for applications requiring programmable or complex shape transformations [6, 7]. For instance, Zhou et al. [7] designed a six-arm structure with multiple locomotion modes, where the arms are magnetized downward, and the main body is magnetized upward. By varying the amplitude and direction of the applied magnetic field, the millirobot can perform tasks, including lying down, 180° flipping, moving to a desired position, and catching and carrying a ceramic ball.

More recently, 3D printing has been exploited for the fabrication of permanent magnets with a desired distribution of the magnetic field [8, 9]. For instance, Kim et al. [8] developed a fabrication technique to program intricate patterns of magnetic polarities in 3D-printed soft materials, which has enabled the use of such h-MREs to create complex modes of transformations. Such a rapid and complex transformation of 3D-printed hard magnetic materials appears to pave the way for new applications in soft robotics, flexible electronics, and biomedical devices.

To deepen the understanding of the behavior of h-MREs and quantitatively predict their complex transformation under actuation, the development of a robust theoretical model and its computational implementation is essential, which has achieved considerable progress in the past several years. For example, Mukherjee et al. [10] proposed a constitutive model that captures the magnetic energy dissipation of embedded hard magnetic particles through ferromagnetic hysteresis, which was further employed to analyze the deformation behavior of an inhomogeneously magnetized cantilever beam. Rambausek et al. [11] developed a nonlinear model to evaluate the magneto-visco-elastic response of h-MREs, detailing its numerical implementation within commercial finite element software. Keip and Sridhar [12] introduced a large-deformation phase-field continuum theory based on a variational principle to analyze micro-magneto-mechanical phenomena in magnetic-particle-filled elastomers by using the theories of micro-magnetic interactions. In contrast to the aforementioned theories, Zhao et al. [13] developed a continuum theory with a simple form, which can describe the complex three-dimensional deformations of h-MREs under an external magnetic field.

This chapter first presents the magnetomechanical coupled model developed by Zhao et al. [13], while a modified version of the magnetic energy is also provided. A comparison between the two models is then presented. The model is further extended by incorporating the glass transition mechanism, which is used to model the magnetic activated shape-memory polymers.

5.2 A Magnetomechanical Coupled Model for h-MREs

5.2.1 Kinematics

The deformation gradient is denoted as $\mathbf{F}$ and its determinant is represented as $J = \det \mathbf{F}$. Let $\mathbf{H}_c$ and $\mathbf{B}_c$ denote the magnetic field vector and the magnetic flux density vector in the current configuration of the considered materials, respectively. In the reference configuration, the nominal magnetic field vector and the magnetic flux density vector are denoted as $\tilde{\mathbf{H}}$ and $\tilde{\mathbf{B}}$, respectively. $\mathbf{H}_c$ and $\tilde{\mathbf{H}}$ are related as

$$\mathbf{H}_c = \mathbf{F}^{-T}\tilde{\mathbf{H}}, \tag{5.1}$$

and the relation between $\mathbf{B}_c$ and $\tilde{\mathbf{B}}$ as

$$\mathbf{B}_c = J^{-1}\mathbf{F}\tilde{\mathbf{B}}. \tag{5.2}$$

In the absence of free current and time variation of the pertinent electromagnetic quantities, the static Maxwell's equations in the current configuration are

$$\text{div}\mathbf{B}_c = 0, \quad \text{curl}\mathbf{H}_c = 0. \tag{5.3}$$

In the reference configuration, the corresponding Maxwell's equations are

$$\text{Div}\tilde{\mathbf{B}} = 0, \quad \text{Curl}\tilde{\mathbf{H}} = 0. \tag{5.4}$$

Hard magnetic materials, once magnetically saturated, maintain a high residual magnetic flux density $\mathbf{B}^r$, even in the absence of applied magnetic fields (Figure 5.1). This behavior is due to their high coercivity, which enables the material to retain this residual magnetic flux density $\mathbf{B}^r$ over a wide range of applied magnetic fields below the coercive field strength $\mathbf{H}^c$. In a hard magnetic soft material, the residual magnetic flux density in the current configuration is defined as

$$\mathbf{B}^r = \mathbf{B}_c|_{\mathbf{H}_c = 0}. \tag{5.5}$$

In the reference configuration, the nominal residual magnetic flux density is defined as

$$\tilde{\mathbf{B}}^r = \tilde{\mathbf{B}}|_{\tilde{\mathbf{H}} = 0}. \tag{5.6}$$

Thus, the relation between $\mathbf{B}^r$ and $\tilde{\mathbf{B}}^r$ can be expressed as

$$\mathbf{B}^r = J^{-1}\mathbf{F}\tilde{\mathbf{B}}^r. \tag{5.7}$$

5.2.2 F-based Model

To facilitate physical interpretation and separation of the magneto-mechanical constitutive laws, the nominal Helmholtz free energy density can be divided into an elastic part as a function of the deformation gradient, and a magnetic part as a function of the deformation gradient and the nominal magnetic flux density as the state parameter. This partitioning allows the elastic component to be associated with any of the widely used constitutive models for soft materials, such as the neo-Hookean, Gent, eight-chain, Ogden models, etc.

For ideal hard magnetic soft materials, the slope of the linear relation between $\mathbf{B}_c$ and $\mathbf{H}$ within the actuation field range is determined by the vacuum (or air) permeability μ_0 as follows:

$$\mathbf{H}_c = \frac{1}{\mu_0}(\mathbf{B}_c - \mathbf{B}^r). \tag{5.8}$$

The magnetic potential energy, or the magnetic part of Helmholtz free energy per unit volume in the current configuration, can be defined as the work required to realign the magnetic moment $\mathbf{B}^r/\mu_0$ along the applied magnetic field $\mathbf{B}^{\text{applied}}$ as follows:

$$W^{\text{magnetic}} = -\frac{1}{\mu_0}\mathbf{B}^r \cdot \mathbf{B}^{\text{applied}}. \tag{5.9}$$

Through considering the relation in Eq. (5.7), the magnetic Helmholtz free energy per unit reference volume can be expressed as

$$\Psi^{\text{magnetic}} = -\frac{1}{\mu_0}\mathbf{F}\tilde{\mathbf{B}}^r \cdot \mathbf{B}^{\text{applied}}. \tag{5.10}$$

The remanent magnetization in the reference configuration can be further defined as $\tilde{\mathbf{M}} = \tilde{\mathbf{B}}^r/\mu_0$.

Once a hard magnetic material is magnetically saturated, the residual magnetic flux density $\tilde{\mathbf{B}}^r$ of the material is fixed unless a strong demagnetizing field beyond the coercivity is applied.

The total Helmholtz free energy of an ideal hard magnetic soft material per unit volume in the reference configuration can be expressed as

$$\Psi = \Psi^{\text{elastic}}(\mathbf{F}) - \frac{1}{\mu_0}\mathbf{F}\tilde{\mathbf{B}}^r \cdot \mathbf{B}^{\text{applied}}. \tag{5.11}$$

The term Ψ depends only on $\mathbf{F}$, as $\tilde{\mathbf{B}}^r$ and $\mathbf{B}^{\text{applied}}$ are specified at each location of interest.

The first Piola–Kirchoff stress can be expressed by

$$\mathbf{P} = \frac{\partial \Psi^{\text{elastic}}(\mathbf{F})}{\partial \mathbf{F}} - \frac{1}{\mu_0}\mathbf{B}^{\text{applied}} \otimes \tilde{\mathbf{B}}^r. \tag{5.12}$$

By converting nominal quantities into true quantities, the Cauchy stress can be obtained as

$$\begin{aligned} \boldsymbol{\sigma} &= \frac{1}{J}\frac{\partial \Psi^{\text{elastic}}(\mathbf{F})}{\partial \mathbf{F}}\mathbf{F}^T - \frac{1}{\mu_0 J}\mathbf{B}^{\text{applied}} \otimes \mathbf{F}\tilde{\mathbf{B}}^r \\ &= \frac{1}{J}\frac{\partial \Psi^{\text{elastic}}(\mathbf{F})}{\partial \mathbf{F}}\mathbf{F}^T - \frac{1}{\mu_0}\mathbf{B}^{\text{applied}} \otimes \mathbf{B}^r. \end{aligned} \tag{5.13}$$

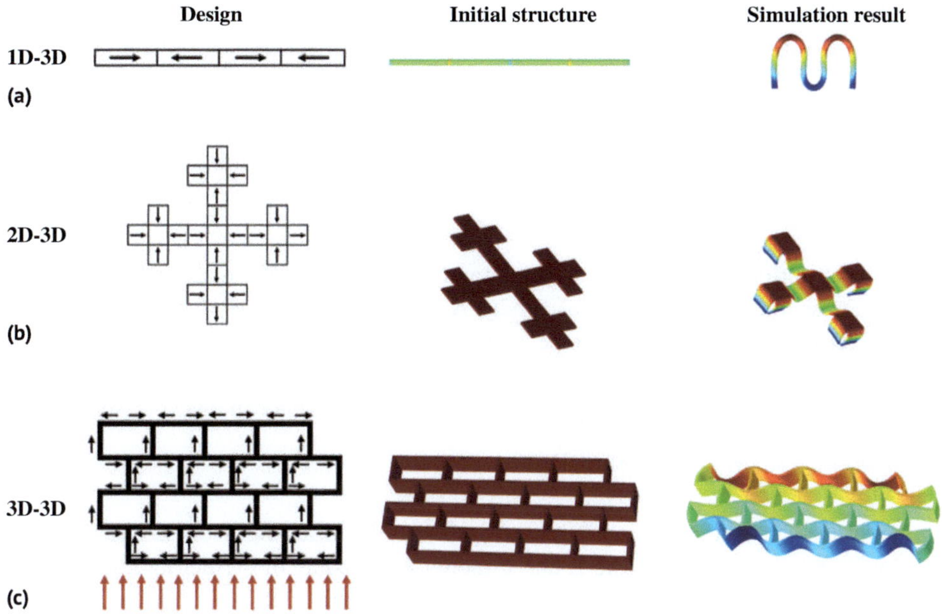

Figure 5.2 Simulation of the shape change in the structures with designed magnetization domain from (a) 1D bar, (b) 2D planar, and (c) 3D structure. The relevant material parameters are chosen as shear modulus $\mu = 330$ kPa, the external magnetic field applied set as 300 mT, and the remanent magnetization set as 81 kA/m. *Source*: Adapted from Kim et al. [8].

To characterize the three-dimensional deformation of h-MREs under an external magnetic field, the mechanical part is assumed to be the quasi-incompressible neo-Hookean model with the bulk modulus three orders of magnitude larger than the shear modulus. The above model was then implemented into finite element software COMSOL through explicitly implementing the above free energy function into the hyperelastic model. As shown in Figure 5.2, the simulations illustrate the evolution of shape from 1D bar and 2D planar shapes to complex 3D structures. The simulated results align with experimental results, confirming that the model effectively predicts complex shape changes facilitated by programmable magnetic domains in 3D-printed structures.

5.2.3 R-based Model

Yan et al. [14] treated the h-MREs as a composite material, where the effective magnetization is attributed to the collective behavior of the dispersed magnetized particles. It is assumed that the effective magnetization of the deformed MRE depends only on the rotation, independent of the stretch, since the individual magnetic particles are too rigid to undergo significant stretching compared to the matrix phase. Thus, this theory modifies the continuum framework established by Zhao et al. [13], specifically to account for the independence of magnetization from the stretching deformation of h-MREs.

Thus, the deformation gradient is first decomposed into the stretch tensor **U** and the rotation tensor **R** as

$$\mathbf{F} = \mathbf{RU}. \tag{5.14}$$

Replacing **F** in Eq. (5.7) with **R** gives

$$\mathbf{B}^{\mathrm{r}} = \frac{1}{J}\mathbf{R}\tilde{\mathbf{B}}^{\mathrm{r}}, \tag{5.15}$$

under the assumption that the matrix is moderately soft, allowing the microscopic particles and the macroscopic h-MRE to rotate affinely. An analytical expression for the inverse of the stretch tensor $\mathbf{U}^{-1}$ as a function of **F** is required for the numerical calculations and is provided in Appendix.

The magnetic Helmholtz free energy per unit reference volume can be expressed as

$$\psi^{\mathrm{magnetic}} = -\frac{1}{\mu_0}\mathbf{R}\tilde{\mathbf{B}}^{\mathrm{r}} \cdot \mathbf{B}^{\mathrm{applied}}. \tag{5.16}$$

For comparison, we refer to this model as the **R**-based model. Both the **F**-based model shown in the previous part and the **R**-based model are capable of predicting the intricate shape transformations of h-MREs under an applied magnetic field.

5.2.4 Comparing R-based and F-based Models

To assess the predictive performance of the **F**-based and **R**-based models in capturing the deformation of h-MRE plates under different configurations of magnetic and mechanical loading, Yan et al. [14] performed several experiments. A square plate fully clamped along its boundaries was subjected to magnetization $\mathbf{M} = \mathrm{M}\mathbf{e}_2$, and a uniform field $\mathbf{B}_\mathrm{a} = \mathrm{B}_\mathrm{a}\mathbf{e}_2$ aligned with **M**. Figure 5.3a, b provides a photograph and schematic illustration of the experimental setup.

Initially, the magnetic fields $\mathbf{B}_\mathrm{a}$ and **M** are oriented in-plane and parallel in the undeformed configuration. A uniform pressure **p** is then applied to the plate, inducing both bending and stretching deformation within the magnetic field. While the magnetic field strength remains constant, the pressure is varied from 0 to 300 Pa. The resulting deflection of the plate generates an angle between $\mathbf{B}_\mathrm{a}$ and **M**, leading to magnetic torques that counteract the deformation.

Figure 5.3c, d plots the normalized deflection at the center of the plate, δ_z/h, as a function of applied pressure **p**, both with and without an external magnetic field (i.e. $\mathbf{B}^\mathrm{a} = \{0, 100\}$ mT). As shown, the experimental data aligns closely with the predictions of the **R**-based model, even at high pressure values. Figure 5.3d shows that in the presence of a magnetic field ($\mathbf{B}^\mathrm{a} = 100$ mT), the **F**-based model deviates considerably from the experimental results, even for small deflections ($\delta_z/h < 1$). This disagreement suggests that the **F**-based model cannot accurately predict the responsive behaviors when the plate deflects with stretching deformation in the presence of an applied field parallel to the magnetization.

Yan et al. [14] further designed several other tests to evaluate the performance of the two models. The predictions from the **R**-based model are in good agreement with experiments, for all the cases, including, the deflection of a suspended plate clamped with varied edges,

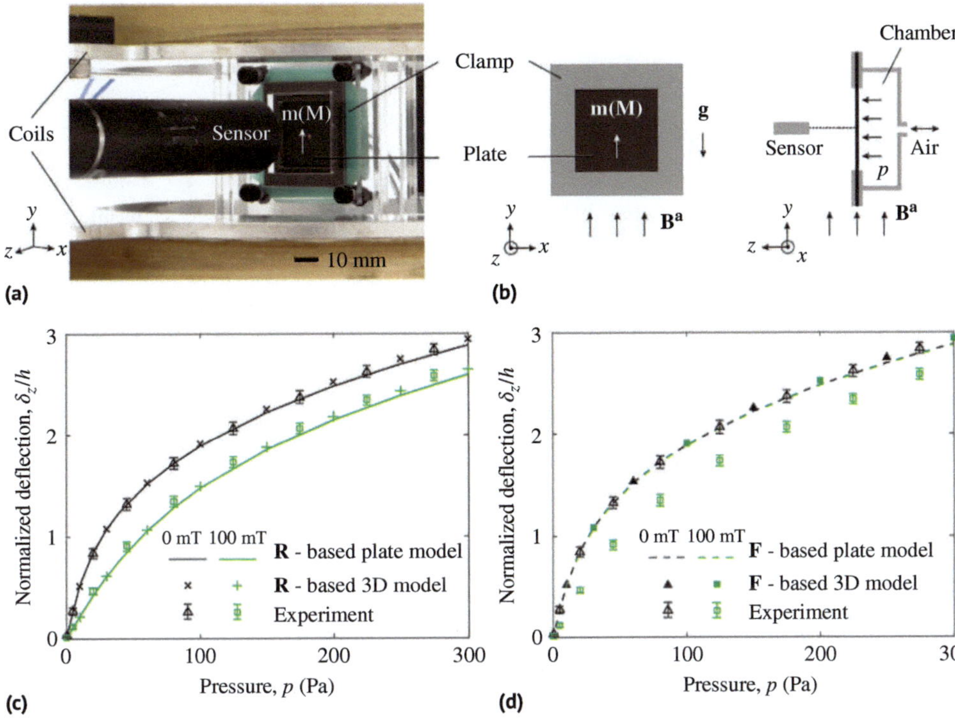

Figure 5.3 Deformation of a fully clamped plate with in-plane magnetization under combined pressure and magnetic loading: (a) photograph of a plate specimen position in between the Helmholtz coils, (b) scheme of the configuration considered in the experiment, (c, d) normalized deflection at the center of the plate, δ_z/h, versus imposed pressure, p, at magnetic flux density $\mathbf{B}^a = \{0, 100\}$ mT. *Source*: Reproduced with permission from Yan et al. [14]/Elsevier/ CC BY-NC-ND 4.0.

deflection of a pressurized, fully clamped plate with in-plane magnetization, deflection of a pressurized, and fully clamped plate with out-of-plane magnetization. In contrast, the **F**-based model fails to accurately predict the deformation of h-MRE plates in scenarios involving significant stretching deformation. This failure is attributed to an incorrect description of the magnetization in the deformed plate.

5.3 Magnetic Activated Shape-memory Polymers

Embedding magnetic particles into the matrix of soft active materials leads to the possibility of achieving multifunctional shape transformations [15–17]. For instance, Sun et al. [15] fabricated ferromagnetic liquid crystal elastomers (LCEs) with nematic order and magnetization, programmable via a customized 3D-printing platform, which can be actuated using either heat sources or external magnetic fields. Tang et al. [16] proposed a thermoresponsive hydrogel incorporating soft magnetic particles, achieving shape changes under a high-frequency alternating magnetic field due to inductive heating. Ze et al. [17] reported magnetic activated shape memory polymers (M-SMPs), which incorporate both soft and

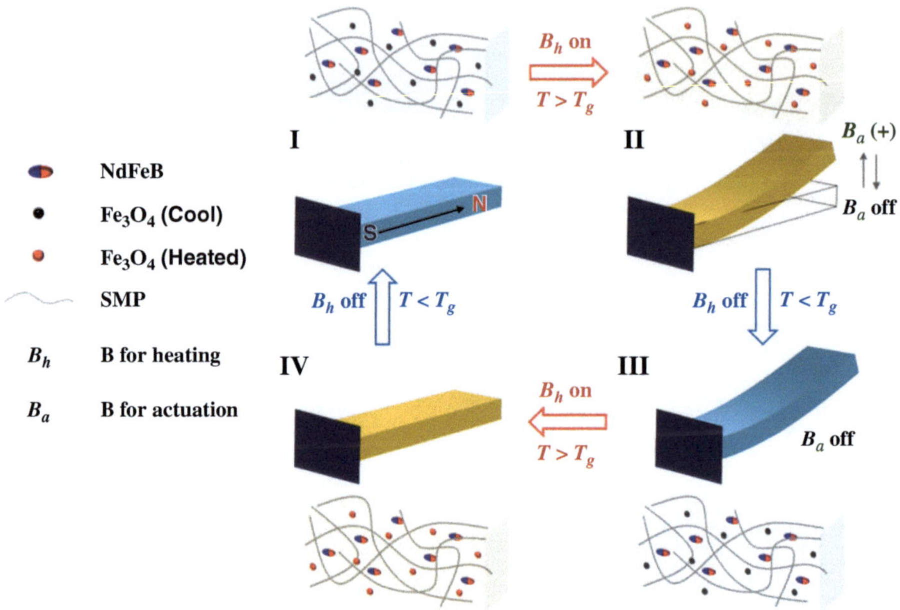

Figure 5.4 The scheme of the shape memory process, which incorporates hard magnetic and soft magnetic particles into the SMP matrix. The temperature of the matrix is increased due to the soft magnetic particles under high frequency AC magnetic field, and the shape deformation of the soft matrix due to the hard magnetic particles under strong external magnetic fields.

hard magnetic particles into the SMP matrix. As shown in Figure 5.4, the matrix can be softened via magnetic inductive heating of low-coercivity particles (stage I), while high-remanence particles with reprogrammable magnetization domains can drive rapid and reversible shape changes under an external magnetic field (stage II). Shape locking can be achieved upon cooling by removing high-frequency alternating current (AC) magnetic fields (stage III). Then, the shape change of the matrix can be recovered by heating soft magnetic particles again (stage IV).

To enable more functional, reliable, and complex shape-morphing structures based on M-SMPs, constitutive models need to be developed to accurately predict the thermo-magnetomechanical coupled process.

Based on the R-based model, we develop a constitutive model for M-SMPs through incorporating the glass transition mechanism, which is described as follows.

5.3.1 Constitutive Theory

For material systems, the first law of thermodynamics is represented as

$$\dot{\mathrm{e}} = \mathbf{P} : \dot{\mathbf{F}} - \nabla_{\mathrm{X}}\mathbf{Q} + \mathrm{R}_{\mathrm{heat}}, \tag{5.17}$$

where e is the internal energy, $\mathbf{P}$ is the first Piola–Kirchhoff stress tensor, $\mathbf{Q}$ is the heat flux, and $\mathrm{R}_{\mathrm{heat}}$ is the heat source.

The entropy inequality can be represented as

$$\dot{\eta} + \nabla_{\mathrm{X}} \cdot \mathbf{H} \geq 0, \tag{5.18}$$

where η is the entropy and $\mathbf{H}$ is the entropy flux. The entropy flux is related to the heat flux as

$$\mathbf{H} = \frac{\mathbf{Q}}{T}. \tag{5.19}$$

The Helmholtz energy density Ψ is defined as

$$\Psi = \mathrm{e} - T\eta, \tag{5.20}$$

where T is temperature.

For M-SMPs, the Helmholtz energy density can be decomposed into four parts as

$$\Psi = \Psi^{\mathrm{T}}(T) + \Psi^{\mathrm{eq}}(\mathbf{F}) + \sum_i \Psi_i^{\mathrm{neq}}(\mathbf{F}, \mathbf{A}_i) + \Psi^{\mathrm{ms}}(\mathbf{F}), \tag{5.21}$$

where Ψ^{T} is the thermal part of free energy density, Ψ^{eq} is the equilibrium mechanical energy density, Ψ_i^{neq} is the ith nonequilibrium mechanical energy density depending on the deformation gradient $\mathbf{F}$ and the internal variable $\mathbf{A}_i$, and Ψ^{ms} is the magnetic part of free energy density.

The rate of Helmholtz free energy density can be represented as

$$\dot{\Psi} = \frac{\partial \Psi^{\mathrm{T}}}{\partial T}\dot{T} + \frac{\partial \Psi^{\mathrm{eq}}}{\partial \mathbf{F}} : \dot{\mathbf{F}} + \sum_i \frac{\partial \Psi_i^{\mathrm{neq}}}{\partial \mathbf{F}} : \dot{\mathbf{F}} + \frac{\partial \Psi^{\mathrm{ms}}}{\partial \mathbf{F}} : \dot{\mathbf{F}} + \sum_i \frac{\partial \Psi_i^{\mathrm{neq}}}{\partial \mathbf{A}_i} : \dot{\mathbf{A}}_i. \tag{5.22}$$

Combining Eqs. (5.17)–(5.19) and (5.22) yields

$$\begin{aligned} &\left[\mathbf{P} - \left(\frac{\partial \Psi^{\mathrm{eq}}}{\partial \mathbf{F}} + \sum_i \frac{\partial \Psi_i^{\mathrm{neq}}}{\partial \mathbf{F}} + \frac{\partial \Psi^{\mathrm{ms}}}{\partial \mathbf{F}}\right)\right] : \dot{\mathbf{F}} - \left(\eta + \frac{\partial \Psi^{\mathrm{T}}}{\partial T}\right)\dot{T} \\ &- \sum_i \frac{\partial \Psi_i^{\mathrm{neq}}}{\partial \mathbf{A}_i} : \dot{\mathbf{A}}_i - \frac{1}{T}\mathbf{Q} \cdot \nabla_{\mathbf{X}} T \geq 0. \end{aligned} \tag{5.23}$$

In the equilibrium state, there is no free energy dissipation, which requires the following relation to be held:

$$\mathbf{P} = \frac{\partial \Psi^{\mathrm{eq}}}{\partial \mathbf{F}} + \sum_i \frac{\partial \Psi_i^{\mathrm{neq}}}{\partial \mathbf{F}} + \frac{\partial \Psi^{\mathrm{ms}}}{\partial \mathbf{F}}, \tag{5.24}$$

$$\eta = -\frac{\partial \Psi^{T}}{\partial T}. \tag{5.25}$$

A stronger form of entropy inequality is then reinforced by requiring that each remaining term in Eq. (5.23) is nonnegative:

$$-\frac{\partial \Psi_i^{\mathrm{neq}}}{\partial \mathbf{A}_i} : \dot{\mathbf{A}}_i \geq 0, \tag{5.26}$$

$$-\frac{1}{T}\mathbf{Q} \cdot \nabla_{\mathbf{X}} T \geq 0. \tag{5.27}$$

The first law of thermodynamics can then be rewritten by substituting Eqs. (5.20), (5.22), (5.24), and (5.25) into Eq. (5.17), which yields

$$T\dot{\eta} = -\nabla_{\mathbf{X}}\mathbf{Q} - \sum_i \frac{\partial \Psi_i^{\mathrm{neq}}}{\partial \mathbf{A}_i} : \dot{\mathbf{A}}_i + \mathrm{R}_{\mathrm{heat}}. \tag{5.28}$$

The thermal part of the free energy density is defined as

$$\Psi^{\mathrm{T}} = c\,(T - T_0) - cT \ln \frac{T}{T_0}, \tag{5.29}$$

where c is the specific heat capacity and T_0 is the reference temperature.

The term Ψ^{eq} is the change in the free energy density due to equilibrium elastic deformation of materials. The neo-Hookean model is used for this term as

$$\Psi^{\mathrm{eq}} = \frac{1}{2} G^{\mathrm{eq}} (\bar{I}_1 - 3) + \frac{1}{2} K (J - 1)^2, \tag{5.30}$$

where $\bar{I}_1$ is the first invariant of the isochoric right Cauchy–Green strain tensor $\bar{\mathbf{C}}$, and G and K are shear and bulk moduli, respectively. It is typically assumed that the bulk modulus is several magnitudes larger than the shear modulus due to incompressibility of soft materials.

We adopt the following form for the nonequilibrium mechanical deformation free energy density Ψ_i^{neq} [18–20]:

$$\Psi_i^{\mathrm{neq}} = \frac{1}{2} G_i^{\mathrm{neq}} \left(\mathbf{A}_i : \bar{\mathbf{C}} - 3 - \ln(\det \mathbf{A}_i)\right), \tag{5.31}$$

where G_i^{neq} is the shear modulus for the ith nonequilibrium process.

The magnetic part of the Helmholtz free energy density Ψ^{ms} is assumed to have the following form [21–23]:

$$\Psi^{\mathrm{ms}} = -\mathbf{R}\tilde{\mathbf{M}} \cdot \mathbf{B}^{\mathrm{applied}}. \tag{5.32}$$

Based on Eqs. (5.24), (5.29)–(5.32), the first Piola stress can be obtained as

$$\begin{aligned} \mathbf{P} = {} & J^{-2/3} G^{\mathrm{eq}} \left[\mathbf{F} - \frac{1}{3} (\mathrm{tr}\mathbf{C})\, \mathbf{F}^{-\mathrm{T}}\right] + JK(J-1)\, \mathbf{F}^{-\mathrm{T}} \\ & + J^{-2/3} \sum_i G_i^{\mathrm{neq}} \mathbf{F} \left(\mathbf{A}_i - \frac{1}{3} (\bar{\mathbf{C}} : \mathbf{A}_i)\, \bar{\mathbf{C}}^{-1}\right) - \left(\frac{\partial \mathbf{R}}{\partial \mathbf{F}}\right)^{\mathrm{T}} (\mathbf{B}^{\mathrm{applied}} \otimes \tilde{\mathbf{M}}), \end{aligned} \tag{5.33}$$

where $\left(\frac{\partial \mathbf{R}}{\partial \mathbf{F}}\right)^{\mathrm{T}}$ is a fourth-order tensor. The calculation of the rotation $\mathbf{R} = \mathbf{R}(\mathbf{F})$ and its derivative $\left(\frac{\partial \mathbf{R}}{\partial \mathbf{F}}\right)^{\mathrm{T}}$ is shown in Appendix [24–26].

The Cauchy stress can be evaluated as

$$\boldsymbol{\sigma} = \frac{1}{J} \mathbf{P}\mathbf{F}^{\mathrm{T}}. \tag{5.34}$$

The corresponding entropy can be obtained by combining Eqs. (5.25) and (5.29) as

$$\eta = c \ln \frac{T}{T_0}. \tag{5.35}$$

The following form is adopted for the evolution equation for $\mathbf{A}_i$ [20]:

$$\dot{\mathbf{A}}_i = \frac{1}{\tau_i} (\bar{\mathbf{C}}^{-1} - \mathbf{A}_i), \tag{5.36}$$

with an initial condition

$$\mathbf{A}_i (\mathbf{X}, t = 0) = \mathbf{I}.$$

Here, τ_i is the relaxation time for the ith relaxation process. This evolution equation shows that each internal variable $\mathbf{A}_i$ approaches the current value of $\bar{\mathbf{C}}^{-1}$ with a relaxation time of τ_i. As shown in Linder et al. [20] and Chapter 2, the above equation satisfies the second law of thermodynamics (Eq. 5.26).

As shown in Chapter 1, the entropy inequality caused by heat conduction can be satisfied using the Fourier law as

$$\mathbf{Q} = -k_0\mathbf{F}^{-1} \cdot \mathrm{Grad}T\mathbf{F}^{-1}, \tag{5.37}$$

where k_0 is the thermal conductivity.

Combining Eqs. (5.28), (5.36), and (5.33) leads to the following form for the first law of thermodynamics:

$$c\dot{T} = k_0\mathbf{F}^{-1} \cdot \mathrm{Grad}T\mathbf{F}^{-1} + \sum_i \frac{G_i^{\mathrm{neq}}}{2\tau_i}[\bar{\mathbf{C}} - \mathbf{A}_i^{-1}] : [\bar{\mathbf{C}}^{-1} - \mathbf{A}_i] + R_{\mathrm{heat}}. \tag{5.38}$$

In the current work, the heat source is generated by soft magnetic particles upon a high-frequency magnetic field. To be simple, we further reinforce a homogeneous temperature field.

5.3.2 Parameter Determination

To quantitatively describe the deformation and mechanical behaviors of M-SMPs, we determine the model parameters using the experimental data in Ze et al. [17].

Here, we treat the magnetic particles and SMP matrix as a uniform material to obtain the corresponding mechanical properties, including the equilibrium modulus, nonequilibrium modulus, and relaxation time.

For the constitutive model, if not considering the magnetic field, the storage modulus and loss modulus upon a small strain dynamic strain can be represented as [27]

$$\begin{aligned} E'_{\mathrm{disc}}(\omega) &= E^{\mathrm{eq}} + \sum_i^N \frac{E_i^{\mathrm{neq}}\omega^2\tau_i^2}{1+\omega^2\tau_i^2}, \\ E''_{\mathrm{disc}}(\omega) &= \sum_i^N \frac{E_i^{\mathrm{neq}}\omega\tau_i}{1+\omega^2\tau_i^2}, \end{aligned} \tag{5.39}$$

where E^{eq} is the equilibrium Young's modulus and E_i^{neq} is the nonequilibrium Young's modulus of the ith relaxation process.

The dynamic temperature sweep test is used to obtain the above parameters. Instead of directly fitting the parameters to the experimental data, we use a continuous spectrum generated from a fractional Zener model [27–29]. The storage modulus and loss modulus can be represented as

$$\begin{aligned} E'_{\mathrm{frac}}(\omega) &= E^{\mathrm{eq}} + \frac{E^{\mathrm{neq}}(\omega\xi)^{2\alpha} + (\omega\xi)^{\alpha}\cos(\alpha\pi/2)}{1+(\omega\xi)^{2\alpha} + (\omega\xi)^{\alpha}\cos(\alpha\pi/2)}, \\ E''_{\mathrm{frac}}(\omega) &= \frac{E^{\mathrm{neq}}(\omega\xi)^{\alpha}\sin(\alpha\pi/2)}{1+(\omega\xi)^{2\alpha} + (\omega\xi)^{\alpha}\cos(\alpha\pi/2)}, \end{aligned} \tag{5.40}$$

where $E^{\text{neq}} = \sum_i^N E_i^{\text{neq}}$ is the total nonequilibrium Young's modulus, ξ is the characteristic stress relaxation time, and $0 \leq \alpha \leq 1$ represents the breadth of the stress relaxation spectrum.

The viscoelastic relaxation spectra $h_{\text{disc}}(\nu)$ and $h_{\text{frac}}(\nu)$ can then be estimated from the complex moduli $E^*(i\omega) = E'(\omega) + iE''(\omega)$ using the inverse Stieltjes transform. The cumulative spectra $H(\nu) = \int_0^\nu h(u)\,du$ for the discrete and fractional models can be further represented as

$$H_{\text{disc}}(\nu) = \sum_i^N E_i^{\text{neq}} \langle \nu - \nu_i \rangle,$$
$$H_{\text{frac}}(\nu) = \frac{E^{\text{neq}}}{\alpha\pi}\left[\arctan\left(\frac{(\lambda\xi)^\alpha + \cos(\alpha\pi)}{\sin(\alpha\pi)}\right) - \pi\left(\frac{1}{2} - \alpha\right)\right], \tag{5.41}$$

where $\langle \nu - \nu_i \rangle = 1$ for $\nu \geq \nu_i$.

A power law distribution for the relaxation frequencies is then assumed:

$$\nu_i^0 = \nu_{\min}^0 \left(\frac{\nu_{\max}^0}{\nu_{\min}^0}\right)^{(j-1)/(N-1)}, \tag{5.42}$$

where $\nu_{\max}^0$ and $\nu_{\min}^0$ are the maximum and minimum of relaxation frequencies.

The nonequilibrium moduli E_i^{neq} are calculated such that $H_{\text{disc}}(\nu)$ forms a step-wise approximation for $H_{\text{frac}}(\nu)$ as

$$E_1^{\text{neq}} = \frac{1}{2}\left(H(\nu_1^0) + H(\nu_2^0)\right),$$
$$E_i^{\text{neq}} = \frac{1}{2}\left(H(\nu_{i+1}^0) - H(\nu_{i-1}^0)\right), 1 < i < N-1, \tag{5.43}$$
$$E_N^{\text{neq}} = E^{\text{neq}} - \sum_i^{N-1} E_i^{\text{neq}}.$$

It is noted that with the increasing values of N, the cumulative spectrum of the discrete model can form a smooth approximation of the continuous model.

The relaxation time depends on temperature. Here, the Williams–Landel–Ferry (WLF) equation is used to represent the shift factor $\mathrm{a}(T)$, which has the following form:

$$\log(\mathrm{a}(T)) = \frac{-C_1(T - T_g)}{C_2 + T - T_g}, \tag{5.44}$$

where C_1 and C_2 are WLF constants at the glass transition temperature T_g. The stress relaxation time is then represented as $\xi = \tau^g \mathrm{a}(T)$ and $\tau_i = \tau_i^g a(T)$.

Equation (5.40) together with the above WLF function is used to fit the storage modulus and loss factor defined as $\tan\delta = E'/E''$. As shown in Figure 5.5, the fractional Zener model can accurately capture the storage modulus of SMP with 15% magnetic particles as a function of temperature. It can also reproduce the main features of the loss factor, including the peak value of tanδ and the glass transition region. The simulated value of the loss factor is smaller than the measured value in the low temperature regime. This is because the WLF function fails to represent the shift factor far below T_g.

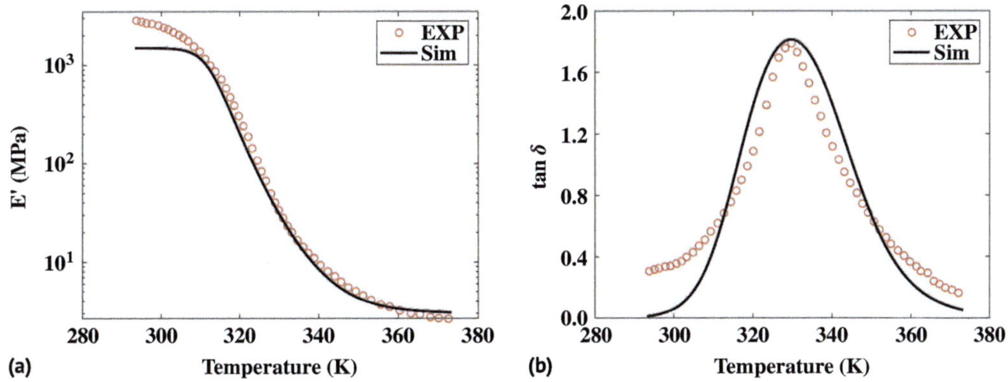

Figure 5.5 Experimental data and model simulation for storage modulus and tan δ as a function of temperature.

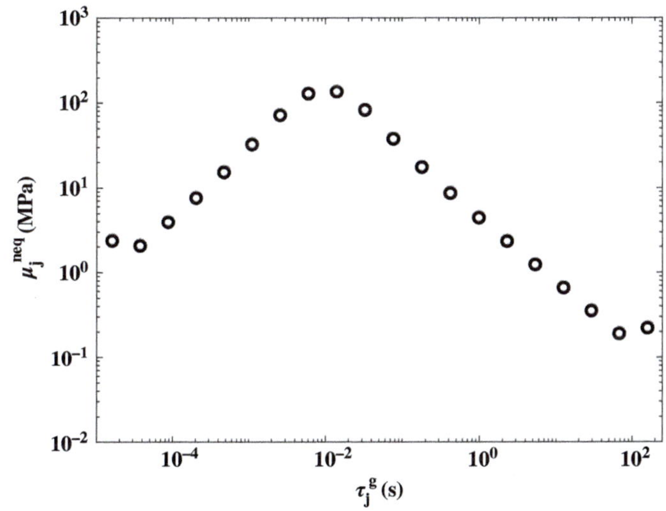

Figure 5.6 The discrete stress spectrum for $N = 20$ relaxation processes.

The shear moduli can then be calculated from Young's modulus using $G^{eq} = E^{eq}/2\,(1+\upsilon_r)$, and $G_i^{neq} = E_i^{neq}/2\,(1+\upsilon_g)$, where the rubbery Poisson's ratio $\upsilon_r = 0.5$ and the glassy Poisson's ratio $\upsilon_g = 0.35$. The corresponding stress relaxation spectrum G_i^{neq}-τ_i^g is plotted in Figure 5.6. The model parameters are listed in Table 5.1.

5.3.3 Simulation Results

To simulate the complete process of M-SMPs (as shown in Figure 5.4), we implement the above model into COMSOL. The external magnetic field is set as a function of time as shown in Figure 5.7 (blue line). The displacement of the M-SMP cantilever is defined by the z-axis displacement at the right end (black line). The simulated value of the displacement

Table 5.1 Model parameters for M-SMPs.

Parameter	Values	Physical significance
G^{eq} (MPa)	1.0	Equilibrium shear modulus
K(MPa)	1666.7	Bulk modulus
C_1	20	The first WLF constant
C_2 (K)	190	The second WLF constant
T_g (K)	323.3	Glass transition temperature
$G^{neq} - \xi$	—	Stress relaxation spectrum
$\lvert\tilde{\mathbf{M}}\rvert$ (kA/m)	100	Remanent magnetization

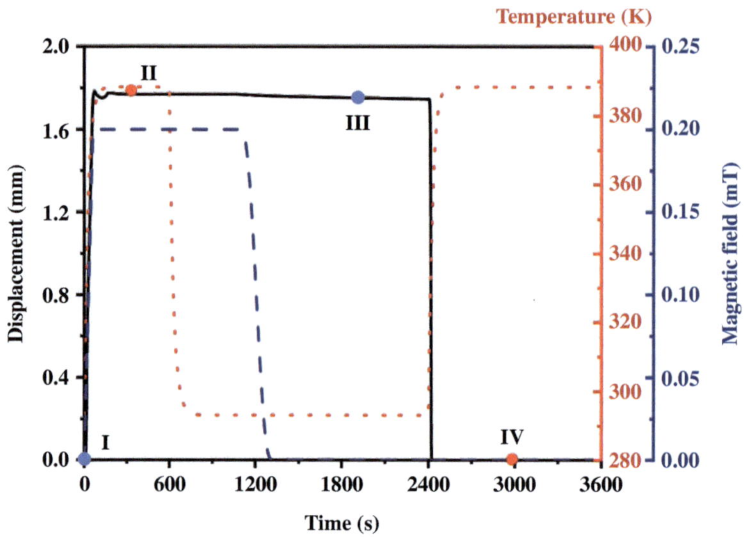

Figure 5.7 The change of the temperature, the strong external magnetic fields, and the displacement of the M-SMP cantilever during complete process.

is consistent with the experimental data as reported in Ze et al. [17]. This suggests that the model is able to capture the whole shape-memory process, including shape programming by the magnetic field, shape locking by cooling and shape recovery through heating. The above model provides a powerful framework for the design of M-SMPs, which combines the unique features of magnetic-activated elastomers and shape-memory polymers.

5.4 Conclusion

In this chapter, we presented a magnetomechanical model for magnetoactive soft materials. The model is able to capture many important responsive features of materials. We also extended the model to describe magnetic activated shape-memory polymers, with the temporary shape programmed by the magnetic field, while the shape-memory effect is the same as in other amorphous polymeric systems. For the current framework, we do not consider the effect of soft magnetic particles. In addition, it is assumed that the residual magnetic flux density does not change upon the applied magnetic field. Recent works [10, 11, 30] have developed a unified framework for MREs comprising magnetically hard or soft magnetic particles, which can be applied for magnetoactive soft materials in various scenarios.

Appendix

(i) To calculate the rotation $\mathbf{R} = \mathbf{R}(\mathbf{F})$, we first compute $\mathbf{U}^{-1}$ explicitly in terms of the invariants of $\mathbf{C} = \mathbf{F}^{\mathrm{T}}\mathbf{F}$, and then compute the rotation using $\mathbf{R} = \mathbf{F}\mathbf{U}^{-1}$. Specifically, with the three invariants of $\mathbf{C}$ denoted by

$$I_1 = \mathrm{tr}\mathbf{C}, I_2 = \frac{1}{2}\left[(\mathrm{tr}\mathbf{C})^2 - \mathrm{tr}\left(\mathbf{C}^2\right)\right], I_3 = \det\mathbf{C}, \tag{5.45}$$

and the three principal invariants of $\mathbf{U}$ denoted by

$$i_1 = \mathrm{tr}\mathbf{U}, i_2 = \frac{1}{2}\left[(\mathrm{tr}\mathbf{U})^2 - \mathrm{tr}\left(\mathbf{U}^2\right)\right], i_3 = \det\mathbf{U}. \tag{5.46}$$

The tensor $\mathbf{U}^{-1}$ may be expressed in terms of the invariants of $\mathbf{C}$ as follows:

$$\mathbf{U}^{-1} = ((i_1 i_2 - i_3)\, i_3)^{-1}\left[\left(i_1 i_2^2 - i_3 i_1^2 - i_3 i_2\right)\mathbf{1} - \left(i_3 + i_1^3 - 2 i_1 i_2\right)\mathbf{C} + i_1 \mathbf{C}^2\right] \tag{5.47}$$

with

$$\begin{aligned} i_1 &= \sqrt{I_1 + 2I_2} \\ i_2 &= \frac{\sqrt{2\sqrt{I_3}\lambda^3 + I_2\lambda^2 - I_3} + \sqrt{I_3}}{\lambda} \\ i_3 &= \sqrt{I_3}, \end{aligned} \tag{5.48}$$

where

$$\lambda = \frac{1}{\sqrt{3}}\left\{I_1 + 2\left[(I_1)^2 - 3I_2\right]^{\frac{1}{2}} \cos\left[\frac{1}{3}\cos^{-1}\left(\frac{2(I_1)^3 - 9I_1 I_2 + 27I_3}{2(I_1 I_1 - 3I_2)^{\frac{3}{2}}}\right)\right]\right\}^{\frac{1}{2}} \tag{5.49}$$

is an eigenvalue of $\mathbf{U}$.

The rotation $\mathbf{R}$ and the stretch tensor $\mathbf{U}$ are then calculated using

$$\mathbf{R} = \mathbf{F}\mathbf{U}^{-1}, \mathbf{U} = \mathbf{R}^{\mathrm{T}}\mathbf{F}. \tag{5.50}$$

(ii) In the expression for the Piola stress, the fourth-order tensor $\mathbf{R} = \mathbf{R}(\mathbf{F})$ acts on the second-order tensor $\mathbf{B}^{\mathrm{app}} \otimes \mathbf{M}_{\mathrm{R}}^{\mathrm{rem}}$. We calculate this term using the formula as

$$\begin{aligned} &\left(\frac{\partial \mathbf{R}}{\partial \mathbf{F}}\right)^{\mathrm{T}}\left[\mathbf{B}^{\mathrm{app}} \otimes \mathbf{m}_{\mathrm{R}}^{\mathrm{rem}}\right] \\ &= (\det\mathbf{Y})^{-1}\mathbf{R}\mathbf{Y}\left(\mathbf{R}^{\mathrm{T}}\left[\mathbf{B}^{\mathrm{app}} \otimes \mathbf{m}_{\mathrm{R}}^{\mathrm{rem}}\right] - \left[\mathbf{B}^{\mathrm{app}} \otimes \mathbf{m}_{\mathrm{R}}^{\mathrm{rem}}\right]^{\mathrm{T}}\mathbf{R}\right)\mathbf{Y}, \end{aligned} \tag{5.51}$$

where

$$\mathbf{Y} \overset{\mathrm{def}}{=} (\mathrm{tr}\mathbf{U})\,\mathbf{I} - \mathbf{U}.$$

References

1 Zvi Rigbi and Leif Jilken. The response of an elastomer filled with soft ferrite to mechanical and magnetic influences. *Journal of Magnetism and Magnetic Materials*, 37(3): 267–276, 1983.

2 Mattias Lokander and Bengt Stenberg. Performance of isotropic magnetorheological rubber materials. *Polymer Testing*, 22(3): 245–251, 2003.

3 Kankanala Danas, Sundeep V Kankanala, and Nicolas Triantafyllidis. Experiments and modeling of iron-particle-filled magnetorheological elastomers. *Journal of the Mechanics and Physics of Solids*, 60(1): 120–138, 2012.

4 Laurence Bodelot, J-P Voropaieff, and Tobias Pössinger. Experimental investigation of the coupled magneto-mechanical response in magnetorheological elastomers. *Experimental Mechanics*, 58: 207–221, 2018.

5 Zhe Chen, Donghao Zhao, Binhong Liu, Guodong Nian, Xiaokeng Li, Jun Yin, Shaoxing Qu, and Wei Yang. 3D printing of multifunctional hydrogels. *Advanced Functional Materials*, 29(20): 1900971, 2019.

6 Monika G Garrell, Bao-Min Ma, Albert J Shih, Edgar Lara-Curzio, and Ronald O Scattergood. Mechanical properties of polyphenylene-sulfide (PPS) bonded Nd–Fe–B permanent magnets. *Materials Science and Engineering: A*, 359(1–2): 375–383, 2003.

7 Ran Zhou, Athira N Surendran, Marcel Mejulu, and Yang Lin. Rapid microfluidic mixer based on ferrofluid and integrated microscale NdFeB-PDMS magnet. *Micromachines*, 11(1): 29, 2019.

8 Yoonho Kim, Hyunwoo Yuk, Ruike Zhao, Shawn A Chester, and Xuanhe Zhao. Printing ferromagnetic domains for untethered fast-transforming soft materials. *Nature*, 558(7709): 274–279, 2018.

9 Anthony P Taylor, Camilo Vélez Cuervo, David P Arnold, and Luis Fernando Velásquez-García. Fully 3D-printed, monolithic, mini magnetic actuators for low-cost, compact systems. *Journal of Microelectromechanical Systems*, 28(3): 481–493, 2019.

10 Dipayan Mukherjee, Matthias Rambausek, and Kostas Danas. An explicit dissipative model for isotropic hard magnetorheological elastomers. *Journal of the Mechanics and Physics of Solids*, 151: 104361, 2021.

11 Matthias Rambausek, Dipayan Mukherjee, and Kostas Danas. A computational framework for magnetically hard and soft viscoelastic magnetorheological elastomers. *Computer Methods in Applied Mechanics and Engineering*, 391: 114500, 2022.

12 Marc-André Keip and Ashish Sridhar. A variationally consistent phase-field approach for micro-magnetic domain evolution at finite deformations. *Journal of the Mechanics and Physics of Solids*, 125: 805–824, 2019.

13 Ruike Zhao, Yoonho Kim, Shawn A Chester, Pradeep Sharma, and Xuanhe Zhao. Mechanics of hard-magnetic soft materials. *Journal of the Mechanics and Physics of Solids*, 124: 244–263, 2019.

14 Dong Yan, Bastien F G Aymon, and Pedro M Reis. A reduced-order, rotation-based model for thin hard-magnetic plates. *Journal of the Mechanics and Physics of Solids*, 170: 105095, 2023.

15 Yuxuan Sun, Liu Wang, Zhengqing Zhu, Xingxiang Li, Hong Sun, Yong Zhao, Chenhui Peng, Ji Liu, Shiwu Zhang, and Mujun Li. A 3D-printed ferromagnetic liquid crystal

elastomer with programmed dual-anisotropy and multi-responsiveness. *Advanced Materials*, 35(45): 2302824, 2023.

16 Jingda Tang, Qianfeng Yin, Yancheng Qiao, and Tiejun Wang. Shape morphing of hydrogels in alternating magnetic field. *ACS Applied Materials & Interfaces*, 11(23): 21194–21200, 2019.

17 Qiji Ze, Xiao Kuang, Shuai Wu, Janet Wong, S Macrae Montgomery, Rundong Zhang, Joshua M Kovitz, Fengyuan Yang, H Jerry Qi, and Ruike Zhao. Magnetic shape memory polymers with integrated multifunctional shape manipulation. *Advanced Materials*, 32(4): 1906657, 2020.

18 Mokarram Hossain, Duc Khoi Vu, and Paul Steinmann. Experimental study and numerical modelling of VHB 4910 polymer. *Computational Materials Science*, 59: 65–74, 2012.

19 Shuolun Wang, Martina Decker, David L Henann, and Shawn A Chester. Modeling of dielectric viscoelastomers with application to electromechanical instabilities. *Journal of the Mechanics and Physics of Solids*, 95: 213–229, 2016.

20 Christian Linder, Mykola Tkachuk, and Christian Miehe. A micromechanically motivated diffusion-based transient network model and its incorporation into finite rubber viscoelasticity. *Journal of the Mechanics and Physics of Solids*, 59(10): 2134–2156, 2011.

21 Mikhail V Vaganov, D Yu Borin, Stefan Odenbach, and Yu L Raikher. Effect of local elasticity of the matrix on magnetization loops of hybrid magnetic elastomers. *Journal of Magnetism and Magnetic Materials*, 459: 92–97, 2018.

22 Mikhail V Vaganov, D Yu Borin, Stefan Odenbach, and Yu L Raikher. Training effect in magnetoactive elastomers due to undermagnetization of magnetically hard filler. *Physica B: Condensed Matter*, 578: 411866, 2020.

23 M Schümann, D Yu Borin, J Morich, and S Odenbach. Reversible and non-reversible motion of NdFeB-particles in magnetorheological elastomers. *Journal of Intelligent Material Systems and Structures*, 32(1): 3–15, 2021.

24 Yi-Chao Chen and Lewis Wheeler. Derivatives of the stretch and rotation tensors. *Journal of Elasticity*, 32: 175–182, 1993.

25 Anne Hoger and Donald E Carlson. Determination of the stretch and rotation in the polar decomposition of the deformation gradient. *Quarterly of Applied Mathematics*, 42(1): 113–117, 1984.

26 Nigel H Scott. $U = C^{1/2}$ and its invariants in terms of C and its invariants. *Journal of Elasticity*, 141(2): 363–379, 2020.

27 Rui Xiao, Jinwoo Choi, Nishant Lakhera, Christopher M Yakacki, Carl P Frick, and Thao D Nguyen. Modeling the glass transition of amorphous networks for shape-memory behavior. *Journal of the Mechanics and Physics of Solids*, 61(7): 1612–1635, 2013.

28 Peter Haupt, Alexander Lion, and E Backhaus. On the dynamic behaviour of polymers under finite strains: constitutive modelling and identification of parameters. *International Journal of Solids and Structures*, 37(26): 3633–3646, 2000.

29 Rui Xiao and Thao D Nguyen. An effective temperature theory for the nonequilibrium behavior of amorphous polymers. *Journal of the Mechanics and Physics of Solids*, 82: 62–81, 2015.

30 Dipayan Mukherjee and Kostas Danas. A unified dual modeling framework for soft and hard magnetorheological elastomers. *International Journal of Solids and Structures*, 257: 111513, 2022.

6

Multi-field Modeling of Liquid Crystal Elastomers

Liquid crystal elastomers (LCEs) are a unique class of materials that combine the elasticity of polymers with the anisotropic properties of liquid crystals. Monodomain LCEs, with liquid crystal mesogens aligned in a preferred direction, exhibit large stimulus-responsive deformation due to the coupling between mesogen rotation and the polymer network. In contrast, polydomain LCEs, characterized by randomly oriented local directors, exhibit isotropic properties macroscopically, while they can also transit to a monodomain state under uniaxial stretching, known as the polydomain-to-monodomain transition. This chapter first introduces the classic theories in describing the phase transition and mechanical behaviors in LCEs. It then presents the recently developed thermal-mechanical-order coupled theories for monodomain LCEs. For polydomain LCEs, a micromechanical model employing internal variables is also introduced, which shows the ability to capture the stress response of polydomain LCEs in multiaxial loading conditions.

6.1 Introduction

Liquid crystal elastomers (LCEs) are a distinctive class of elastomers that exhibit elasticity in both isotropic and liquid crystalline states. LCEs are formed through incorporating ordered liquid crystal molecules into polymer chains [1, 2]. LCEs can be categorized into monodomain and polydomain states based on the global alignment of liquid crystal monomers. By orienting the liquid crystal mesogens in a specific direction, monodomain LCEs are obtained, exhibiting stimuli-responsive deformation due to the coupling of liquid crystal monomer rotation with the polymer network [3–5]. In contrast, polydomain LCEs lack a global director but instead possess local directors within each domain, resulting in isotropic properties macroscopically [6, 7].

The pioneering work of Finkelmann et al. [8] reported the synthesis of monodomain LCEs using a secondary cross-linking technique. There has been tremendous progress in synthesizing LCEs in the past decades. The preparation methods of monodomain LCEs can be divided into two categories [9]. The first is the two-step cross-linking approach, where pre-cross-linked liquid crystal polymers are oriented using techniques such as mechanical stretching, field induction, or 3D printing, followed by additional cross-linking to lock this alignment. The second is the one-step cross-linking approach, where monodomain LCEs are fabricated by directly polymerizing low-molecular-weight liquid crystal monomers that

Multi-field Modeling of Soft Active Materials: Properties and Design, First Edition. Rui Xiao.

are pre-aligned. In the absence of meticulous control during cross-linking, the resultant elastomers in the nematic state exhibit a polydomain structure, characterized by domains that are randomly oriented. Upon subjecting the material to uniaxial stretch, the directors within the polydomain LCEs reorient to align along a macroscopic director axis, leading to a transition to a fully transparent state after reaching a critical level of extension, known as polydomain-to-monodomain (P-M) transition [8, 10, 11]. This transformation is also signified by the appearance of a stress plateau in the stress–strain curve [12].

As temperature increases, monodomain LCEs can transit into an isotropic phase, known as the nematic-isotropic (N-I) phase transition. The Landau-de Gennes theory is the most widely used theoretical framework for describing the N-I phase transition [13], while phenomenological functions are also adopted to model the thermo-order coupling behaviors of monodomain LCEs, focusing on the stretch–temperature relationship obtained from experiments to predict shape changes [14, 15]. Upon external stress, LCEs also exhibit complex mechanical-order coupling behavior. Bladon et al. [16] introduced a neoclassical model by incorporating anisotropy into the neo-Hookean model. Schätzle et al. [17] utilized the coupling between the Landau-de Gennes energy, expressed as an expansion in both the small order parameter and strain, to describe the rise in the transition temperature T_{NI} under external stress applied along the nematic alignment direction. Finkelmann et al. [18] incorporated the Landau-de Gennes energy together with neoclassical free energy, taking into account the biaxial effect to interpret the anisotropic behavior of Young's modulus. Jin et al. [19] developed a thermomechanical framework for thermomechanical-order coupling in LCEs. Wu et al. [20] developed a phenomenological model for nematic LCEs by incorporating a triaxial order tensor and an evolution condition, effectively capturing the dependence of thermomechanical responses on loading history.

Continuum theories have been developed and applied to explore unique stress and strain concentration behaviors in LCEs. These theories successfully replicate stripe domain instabilities observed in perpendicular extension experiments on monodomain LCEs [21–24]. For instance, Brighenti et al. [25] presented a micromechanical model integrating the evolution of the chain distribution tensor into LCE networks. Additionally, several models [26–28] can describe stress responses under arbitrary loading directions relative to mesogen orientations.

A few models have also been developed for polydomain LCEs. Fridrikh and Terentjev [29] introduced models of random anisotropy to characterize the polydomain state. Biggins et al. [30] adopted neoclassical and semi-soft free energies for each nematic domain, providing macroscopic behavior predictions for various types of polydomain LCEs. Whitmer et al. [10] employed a coarse-grained model and molecular dynamics simulations to explore the P-M transition under external loading in polydomain LCEs. Zhou and Bhattacharya [31] developed a computational model to understand the coupling between microstructural evolution and macroscopic responses in polydomain LCEs. Lee et al. [32] proposed a macroscale constitutive model for polydomain LCEs, based on insights from micromechanical models, which accurately replicates the experimentally observed responses under various multiaxial loading conditions.

This chapter starts with introducing the general theory of LCEs, including the Landau-de Gennes theory and the neoclassical model. The constitutive model for monodomain LCEs is then introduced for both considering and not considering the orientation of directors. Finally, the current development on modeling polydomain LCEs is presented.

6.2 General Theory for Liquid Crystal Elastomers

6.2.1 Liquid Crystal

Liquid crystal molecules are usually long, rod shaped, or disk shaped. Liquid crystals can be categorized based on their different orientations and positional ordering. This chapter focuses on the most common nematic phase liquid crystals. In nematic liquid crystals, molecules have a certain average alignment direction, while the position of each molecule is disordered. This alignment direction can be described by a unit vector, known as the director, denoted as **n**. As shown in Figure 6.1, we can use the spherical coordinates θ, ϕ to describe the orientation of a rod. The rod-shaped liquid crystal monomers deviate slightly from the director **n** due to thermal motion. The angle between the rod's axis and director, caused by this thermal motion, is defined as θ, which obviously ranges from 0 to $\pi/2$.

Using statistical methods, the degree of ordering of the liquid crystal molecules can be quantified by the order parameter Q:

$$Q = \left\langle \frac{3}{2}\cos^2\theta - \frac{1}{2} \right\rangle, \tag{6.1}$$

where $\langle \cdots \rangle$ denotes an average over all mesogens directions. The value of Q ranges from -0.5 to 1. For a perfectly ordered nematic phase, $Q = 1$, and for an isotropic phase with randomly distributed mesogens, $Q = 0$. A value of $Q = -1/2$ indicates that all the rods are confined to a plane perpendicular to the director **n** [33]. If the coordinate axis z is chosen along **n**, an order tensor can be introduced to describe both the direction and the deviation as follows:

$$\mathbf{Q} = \frac{Q}{2}(3\mathbf{n} \otimes \mathbf{n} - \mathbf{I}). \tag{6.2}$$

The order parameter tensor is traceless. In this case, we consider only the uniaxial situation, with all angles ϕ of rod-like monomers being equivalent. If all directions are perpendicular to **n**, the refractive index should remain the same. When this assumption is not satisfied under external fields, the biaxiality needs to be introduced to form a more complex order parameter. The biaxial order b can be defined as follows:

$$b = \frac{3}{2}\left\langle \sin^2\theta \cos 2\phi \right\rangle. \tag{6.3}$$

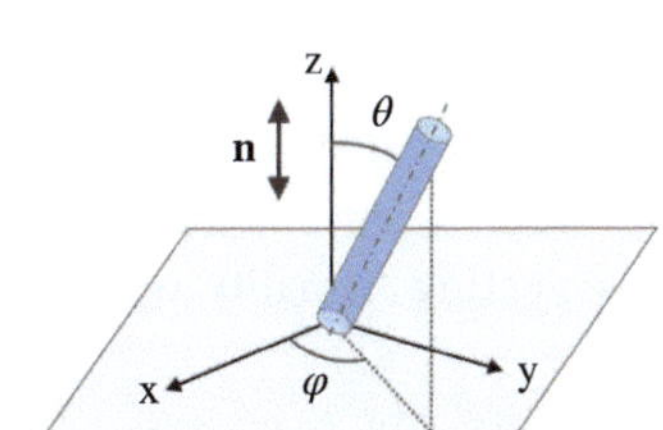

Figure 6.1 The coordinates of a rod used to define the order parameter tensor.

When it is symmetric about **n**, then b will be zero since $\langle \cos 2\phi \rangle = 0$. When the director **n** is along z, the order tensor for the biaxial state can be rewritten as

$$\mathbf{Q} = \begin{pmatrix} -(Q-b)/2 & 0 & 0 \\ 0 & -(Q+b)/2 & 0 \\ 0 & 0 & Q \end{pmatrix}. \tag{6.4}$$

Experimental data indicated that the value of order parameter significantly depends on temperature. When the temperature reaches a certain threshold, the value of Q undergoes a shift from $0.25 \sim 0.5$ to 0. This phenomenon is caused by a phase transition. The critical temperature at which this occurs is called the N-I transition temperature, T_{NI}. In addition to thermal fields, certain specialized liquid crystals can change their ordering or pointing directions when exposed to electric, magnetic, or optical fields. During the N-I phase transition, the liquid crystals lose their directional ordering, although the changes in the thermodynamic parameters are negligible [33].

Landau-de Gennes developed a phase-transition model to elucidate the N-I transition of liquid crystals. This model has become a standard and widely used framework for describing the relationship between the order parameter Q and temperature.

The Landau-de Gennes expansion of free energy in terms of the tensor order $\mathbf{Q}$ is given as follows:

$$\Psi_{nem} = f_0(\mathbf{T}) + \frac{1}{3}ATr(\mathbf{Q}^2) - \frac{4}{9}BTr(\mathbf{Q}^3) + \frac{2}{9}CTr(\mathbf{Q}^4) + \cdots, \tag{6.5}$$

where $A, B, C, \cdots$ are constants. Upon substituting Eq. (6.4) into Eq. (6.5), we can obtain the free energy density in the biaxial case

$$\Psi_{nem} = \frac{1}{2}AQ^2 - \frac{1}{3}BQ^3 + \frac{1}{4}CQ^4 + \cdots + \frac{1}{2}\left(\frac{A}{3} + \frac{2BQ}{3} + \frac{CQ^2}{3} \cdots\right)b^2 + \cdots. \tag{6.6}$$

The Ψ_{nem} here is a function of order parameter, temperature, and biaxiality. The stable phase exists such that the free energy is minimized, i.e. satisfies

$$\frac{\partial \Psi_{nem}}{\partial Q} = AQ - BQ^2 + CQ^3 + \frac{1}{3}b^2(B + CQ) = 0, \tag{6.7a}$$

$$\frac{\partial \Psi_{nem}}{\partial b} = \left(\frac{A}{3} + \frac{2BQ}{3} + \frac{CQ^2}{3}\right)b = 0. \tag{6.7b}$$

The biaxial component of the system is expanded only up to b. This is because the Landau energy is primarily used to evaluate the penalty for inducing b in an otherwise uniaxial phase, where b is equal to zero.

The most sensitive temperature behavior can be captured by rewriting A as $A = A_0(T - T^*)$, while T^* is the lowest temperature at which an isotropic phase can be stabilized. When T is increased to T^*, the minimum in Ψ_{nem} at $Q = 0$ becomes a maximum. Neglecting biaxiality, we arrive at three possible solutions for Q

$$Q_m^0 = 0, \quad Q_m^{\pm}(T) = \frac{B}{2C}\left(1 \pm \sqrt{1 - \frac{4A_0(T - T^*)}{B^2/C}}\right). \tag{6.8}$$

The solution Q_m^- lacks a physical interpretation, as the energy it represents is consistently greater than that of Q_m^+. The two remaining solutions, Q_m^0 and Q_m^+ correspond to the nematic and the isotropic phase, respectively. The condition for the existence of a real solution for Q_m^+ is given by

$$T < T^* + \frac{B^2}{4AC}. \tag{6.9}$$

When the energies of the two stable phases are equal, a phase transition occurs. Consequently, the transition temperature T_{NI} can be calculated as follows:

$$T_{\text{NI}} = T^* + \frac{2B^2}{9AC}. \tag{6.10}$$

6.2.2 Nematic Polymers

Nematic polymers are materials that combine liquid crystals with polymer chains. Cross-linking with polymer chains restricts the free rotation of liquid crystals, resulting in novel properties that arise from the interaction and mutual restriction of two components. Nematic polymers are the building blocks of the LCEs. The classification of nematic polymers is based on the cross-linking position of the liquid crystal monomers attached to the polymer backbone. This leads to the identification of two main categories: main chain and side chain nematic polymers.

In main-chain nematic polymers, the liquid crystal monomers are integrated directly into the polymer backbone, making the liquid crystal alignment closely tied to the backbone. This results in strong anisotropic behavior, as both the mechanical and liquid crystalline properties are closely coupled. In contrast, side-chain nematic polymers feature liquid crystal monomers attached to the polymer backbone through flexible spacers. This configuration allows the liquid crystal molecules more freedom to reorient independently of the backbone, offering more flexibility in tuning their properties, especially under external fields such as electric or magnetic fields.

6.2.2.1 Step Length Tensor

Liquid crystals combine with polymer chains to form nematic polymers, where the average length of chain segments can vary depending on the direction. Warner and Terentjev [33] described the shape of the polymer backbone, accounting for the presence of liquid crystals, through the step tensor ℓ

$$\ell = \frac{3}{aN}\langle \mathbf{R} \otimes \mathbf{R} \rangle = \frac{3a}{N}\sum_{i,j}^{N} \langle \boldsymbol{r}_i \otimes \boldsymbol{r}_j \rangle, \tag{6.11}$$

where $\mathbf{R}$ represents the distance between the head and tail of a polymer chain, a is the average length of the liquid crystal monomers, and N is the average number of segments of the polymer segments. Furthermore, $\mathbf{r}_i$ represents the orientation direction of the ith liquid crystal monomer.

The eigenvalue of the step tensor represents the equivalent length of the main axis of the polymer chain of the liquid crystal. Neglecting the biaxiality, the effective step length tensor can be written as

$$\ell = \ell_{\perp}\mathbf{I} + (\ell_{\parallel} - \ell_{\perp})\,\mathbf{n}\otimes\mathbf{n}, \tag{6.12}$$

where $\ell_{\parallel}$ and $\ell_{\perp}$ are the effective lengths of steps in the directions parallel and perpendicular to $\mathbf{n}$, respectively. $\mathbf{I}$ is a 3-by-3 identity matrix. The step length tensor is temperature dependent. Due to temperature-ordering coupling, there is a direct relationship between components of ℓ and the order parameter Q. In the reference frame where $\mathbf{n}$ is aligned with the z-axis, the tensor takes the following form:

$$\ell = a\begin{pmatrix} 1-Q & 0 & 0 \\ 0 & 1-Q & 0 \\ 0 & 0 & 1+2Q \end{pmatrix}. \tag{6.13}$$

The degree of anisotropy in the nematic polymer can thus be measured by

$$r = \frac{\ell_{\parallel}}{\ell_{\perp}} = \frac{1+2Q}{1-Q}. \tag{6.14}$$

Macroscopically, polydomain LCEs behave isotropically. In contrast, monodomain LCEs have a globally aligned director $\mathbf{n}$, with liquid crystal molecules uniformly oriented in a single direction. As a result, they exhibit significant anisotropy in both mechanical response and thermal deformation properties.

6.2.3 Neoclassical Model

As shown in Figure 6.2, we consider the configuration of the LCEs in a stress-free, low-temperature nematic phase as the reference configuration $\hat{\mathcal{D}}_0$. We define ℓ_0 and ℓ as step length tensors in the reference and current configurations, respectively. The term $\ell_0^{1/2}$ refers to the spontaneous deformation of the elastomer induced by the rotation of the liquid crystal mesogens during the transition from the high-temperature isotropic configuration $\mathcal{D}_i$ to the reference configuration $\hat{\mathcal{D}}_0$. Similarly, $\ell^{1/2}$ represents the spontaneous deformation from the specific deformed configuration $\mathcal{D}_N$ to the current configuration $\mathcal{D}$, driven by changes in nematic order. We further define the deformation gradient from $\mathcal{D}_i$ to $\mathcal{D}_N$ as $\mathbf{F}_N$. Thus, we have

$$\ell^{1/2}\mathbf{F}_N = \mathbf{F}\ell_0^{1/2}. \tag{6.15}$$

This equation can be rewritten as

$$\mathbf{F}_N = \ell^{-1/2}\mathbf{F}\ell_0^{1/2}. \tag{6.16}$$

Considering $\ell_0^{1/2}$ and $\ell^{1/2}$ are spontaneous deformations, the mechanical deformation of the network that stores elastic energy is $\mathbf{F}_N$, which is the deformation gradient of the network. When the network is not mechanically deformed, $\mathbf{F}_N = \mathbf{I}$, $\mathbf{F}$ simplifies to $\mathbf{F} = \ell^{1/2}\ell_0^{-1/2}$, which corresponds to the stress-free stretch tensor resulting from the rotation of the director from $\mathbf{n_0}$ to $\mathbf{n}$.

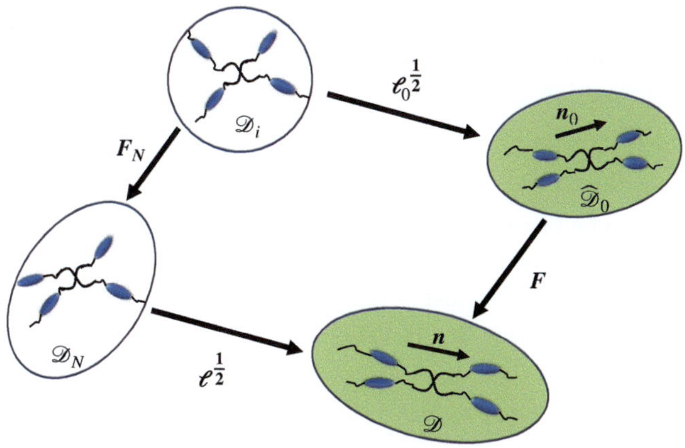

Figure 6.2 Schematic of deformation maps illustrating the reference configuration $\hat{\mathcal{D}}_0$, the current configuration $\mathcal{D}$, the undeformed isotropic configuration $\mathcal{D}_i$, the intermediate configurations $\mathcal{D}_N$, and the deformation gradients between these configurations.

In classic hyperelastic models, the free energy density depends on the invariants of the Cauchy–Green deformation tensor. For example, the free energy density of the neo-Hookean model is scaled with I_1, where $I_1 = \mathrm{Tr}(\mathbf{C})$ and $\mathbf{C} = \mathbf{F}^{\mathrm{T}}\mathbf{F}$. In the same way, we can express the free energy density of the LCEs by $\mathbf{F}^N$ as

$$\Psi_{\mathrm{NC}} = \frac{G}{2} I_{N1}, \tag{6.17}$$

where G is shear modulus, $\mathbf{C}_N = \mathbf{F}_N^{\mathrm{T}}\mathbf{F}_N$, $I_{N1} = \mathrm{Tr}(\mathbf{C}_N)$. This is the neoclassical free energy density developed by Bladon et al. [16].

Similarly, we can exploit other hyperelastic models to obtain the free energy density for LCEs, such as the Mooney–Rivlin model [34], the Ogden-type model [35], and the Gent model [36].

6.3 Thermomechanical Coupled Model for Monodomain Liquid Crystal Elastomers

Monodomain LCEs have been widely employed for designing active structures due to their shape change behaviors with temperature. Thus, it is necessary to construct a simple thermomechanical coupled model with the ability to simulate the shape-morphing behaviors of structures based on monodomain LCEs.

The free energy of monodomain LCEs can be represented by the neoclassical model as

$$\Psi = \frac{G}{2}\left[\mathrm{Tr}\left(\ell_0 \cdot \mathbf{F}^{\mathrm{T}} \cdot \ell^{-1} \cdot \mathbf{F}\right) - 3\right] + \frac{\kappa}{2}(J-1)^2 - G\ln J, \tag{6.18}$$

where κ is the bulk modulus, typically several orders large of the shear modulus.

The principal value of ℓ is described as a function of the order parameter Q according to Eq. (6.13). The current step length tensor is calculated by Eq. (6.12). The $\mathbf{n}$ can be computed from the initial director and the deformation gradient tensor as

$$\mathbf{n} = \frac{\mathbf{Fn}_0}{|\mathbf{Fn}_0|}. \tag{6.19}$$

In a Cartesian coordinate system, it can be expressed in terms of the azimuthal angle in the xy-plane and the polar angle relative to the positive z-axis.

The order parameter Q can be further simplified just as temperature dependent. The following phenomenological function is adopted for $Q(T)$ as [14]

$$Q(T) = \left[1 + \exp^{\left(\frac{(T-T_{\mathrm{ni}})}{c}\right)}\right]^{-1}, \tag{6.20}$$

where c is a constant material-dependent parameter. Apparently, when T is well above T_{NI}, Q approaches zero representing an isotropic state. While T is well below T_{NI}, Q approaches 1, representing a nematic state.

The aforementioned LCE model can be implemented in COMSOL for finite element simulations. The free energy functions and corresponding parameters are defined within the Solid Mechanics Module. The parameters are set as $G = 1$ MPa, $\kappa = 100$ MPa, $T_{\mathrm{NI}} =$ 353.15 K, and $c = 22$.

A bilayer sheet of monodomain LCEs is constructed. The thickness of the bilayer is considerably less than the width and length. The director of the upper layer is aligned in the x-direction, while that of the bottom layer is aligned in the z-direction. When subjected to heating, the bilayer exhibits a predominant contraction along the vertical direction, resulting in incompatible out-of-plane deformation. The ratio of the length (l) to the width (w) of the sheet affects the morphing shape. When $l/w = 1$, the deformation is saddle-shaped as shown in Figure 6.3a. Conversely, when $l/w \gg 1$, the double layer produces a bending deformation as shown in Figure 6.3b.

The anisotropic deformation response of LCEs provides flexibility in designing out-of-plane deformation. Similar to the previous case, a bilayer strip model can be constructed

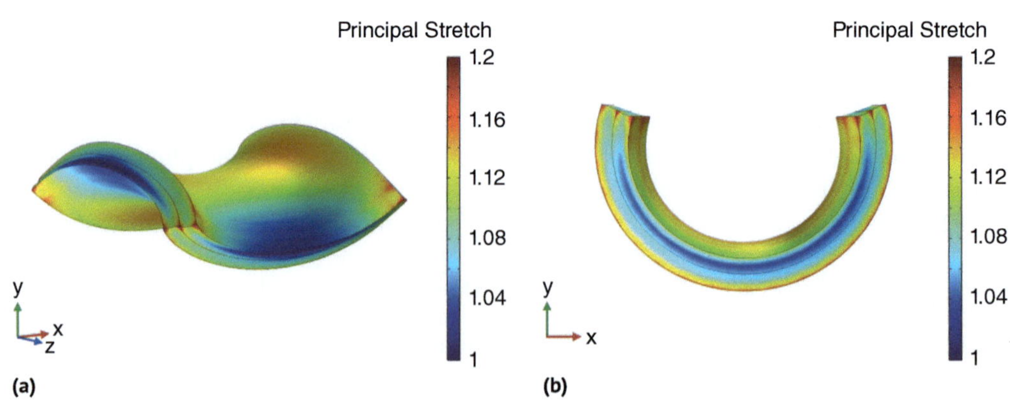

Figure 6.3 Deformation of a bilayer LCE sheet heated to 353.15 K, (a) $l/w = 1$ and (b) $l/w = 4$.

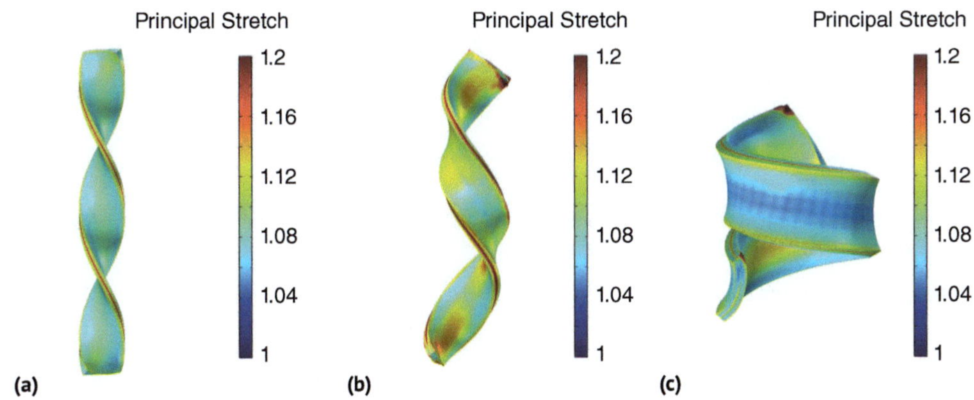

Figure 6.4 Helical deformation of bilayer LCE strips heated to 353.15 K with alignment angles of (a) 45°, (b) 30°, and (c) 15° relative to the long edge.

where the directors of the two layers are perpendicular to each other but angled relative to the long edge of the rectangular sheet. As illustrated in Figure 6.4, the upper LCE layers are aligned at angles of 45°, 30°, and 15° to the long edge. The thermal responses form distinct helical shapes. With the same bilayer strip geometry, a smaller angle to the long edge results in a tighter helix, with a larger radius and a smaller pitch.

6.4 A Viscoelastic Micropolar Theory for Monodomain Liquid Crystal Elastomers

In LCEs, the liquid crystal mesogens undergo gradual rotation due to viscosity. Wang et al. [36] developed a viscoelastic theory for monodomain LCEs, which incorporates the effects of director rotation, along with an independent generalized description of the viscoelastic behavior of the elastomeric network. A complete thermodynamic framework has been provided in Wang et al. [36]. Here, we briefly summarize the derivative process and the governing equations.

The rheological model shown in Figure 6.5 illustrates the interaction of network deformation and director rotation. On the right side, a standard linear solid model is used to represent the viscoelastic behavior of the network. The standard linear solid model (also known as the Zener model) consists of an entropic spring, which accounts for the equilibrium elastic behavior, in parallel with a Maxwell element that captures the time-dependent, nonequilibrium behavior. On the other side, a dashpot is parallel to the stress-free rotation of the director, to represent the viscous resistance to director rotation. Initially, the dashpot inhibits director rotation, causing the deformation of the network element and generating an initial viscoelastic stress response. As the dashpot relaxes, the director begins to rotate, leading to a softening of the stress response from the peak stress achieved by the network to the flow stress, determined by the strain rate and director viscosity. Once the director has rotated fully to the loading direction, further loading deforms the network, producing a viscoelastic stress response.

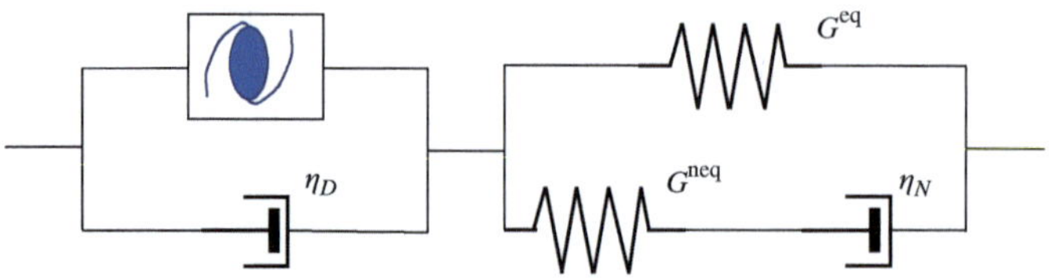

Figure 6.5 Rheological diagrams to conceptually illustrate the interaction of network deformation and director rotation for the monodomain LCEs.

6.4.1 Kinematics

The micropolar continua theory is adopted for the basis of thermodynamics [37]. For a micropolar theory, the deformation field is still described by a deformation gradient $\mathbf{F}$ mapping the point vector $\mathbf{X}$ to $\mathbf{x}(\mathbf{X})$. A director field $\mathbf{n}$ is also defined. A body force $\mathbf{b}$ and a surface traction $\mathbf{t}$ are directly related to this mechanical deformation. The director $\mathbf{n}^0$ in the reference configuration can be mapped to that in the current configuration through a rotation field $\mathbf{R}$ as $\mathbf{n} = \mathbf{R}\mathbf{n}^0$. A director-force $\boldsymbol{\beta}$ and a surface director-traction $\mathbf{s}$ are further defined to be associated with the director field $\mathbf{n}$.

To model the viscoelastic behavior of the material body, we introduce a decomposition of the deformation gradient $\mathbf{F}$ into a multiplicative form

$$\mathbf{F} = \mathbf{F}^{e}\mathbf{F}^{v}. \tag{6.21}$$

Here, $\mathbf{F}^e$ represents the elastic part of the deformation gradient and $\mathbf{F}^v$ denotes the viscous part. The viscous component $\mathbf{F}^v$ is considered an internal state variable that characterizes the material's nonequilibrium response.

6.4.2 Balance of Linear and Angular Momentum

The external power acting on a deformed body can be represented as

$$P_{\text{ext}}(\mathbf{v}, \dot{\mathbf{n}}) = \int_{\partial\mathcal{P}} (\mathbf{t}\cdot\mathbf{v} + \mathbf{s}\cdot\dot{\mathbf{n}})\,\mathrm{d}a + \int_{\mathcal{P}} (\mathbf{b}\cdot\mathbf{v} + \boldsymbol{\beta}\cdot\dot{\mathbf{n}})\,\mathrm{d}v, \tag{6.22}$$

where $\mathbf{v}$ is the velocity field.

Consider a change of observer using the relations $\mathbf{x}^* = \mathbf{Q}\mathbf{x} + \mathbf{y}$ and $\mathbf{n}^* = \mathbf{Q}\mathbf{n}$, where $\mathbf{Q}$ is a second-order tensor and $\mathbf{y}$ is a vector. Thus, the external power can then be represented as

$$P^*_{\text{ext}}(\mathbf{v}^*, \dot{\mathbf{n}}^*) = \int_{\partial\mathcal{P}} (\mathbf{t}^*\cdot\mathbf{v}^* + \mathbf{s}^*\cdot\dot{\mathbf{n}}^*)\,\mathrm{d}a^* + \int_{\mathcal{P}} (\mathbf{b}^*\cdot\mathbf{v}^* + \boldsymbol{\beta}^*\cdot\dot{\mathbf{n}}^*)\,\mathrm{d}v^* \tag{6.23}$$

Considering that the force vectors follow the rules: $\mathbf{t}^* = \mathbf{Q}\mathbf{t}$, $\mathbf{b}^* = \mathbf{Q}\mathbf{b}$, $\mathbf{s}^* = \mathbf{Q}\mathbf{s}$ and $\boldsymbol{\beta}^* = \mathbf{Q}\boldsymbol{\beta}$, the preceding equation can be further rewritten as

$$\begin{aligned} P^*_{\text{ext}}(\mathbf{v}^*, \dot{\mathbf{n}}^*) = {} & P_{\text{ext}}(\mathbf{v}, \dot{\mathbf{n}}) + \alpha\cdot\Big(\int_{\partial\mathcal{P}} \mathbf{t}\,\mathrm{d}a + \int_{\mathcal{P}} \mathbf{b}\,\mathrm{d}v\Big) \\ & + \lambda\cdot\Big(\int_{\partial\mathcal{P}} (\mathbf{x}\times\mathbf{t} + \mathbf{n}\times\mathbf{s})\,\mathrm{d}a + \int_{\mathcal{P}} (\mathbf{x}\times\mathbf{b} + \mathbf{n}\times\boldsymbol{\beta})\,\mathrm{d}v\Big), \end{aligned}$$

where λ is the axial vector of $\mathbf{Q}^{\mathrm{T}}\dot{\mathbf{Q}}$ and $\alpha = \mathbf{Q}^{\mathrm{T}}\dot{\mathbf{y}}$.

The external power should be independent of the observer, which gives

$$P^*_{\text{ext}}(\mathbf{v}^*, \dot{\mathbf{n}}^*) = P_{\text{ext}}(\mathbf{v}, \dot{\mathbf{n}}). \tag{6.24}$$

Equation (6.24) should hold for any observer. This further gives

$$\int_{\partial\mathcal{P}} \mathbf{t}\, \mathrm{d}a + \int_{\mathcal{P}} \mathbf{b}\, \mathrm{d}v = \mathbf{0}, \tag{6.25a}$$

$$\int_{\partial\mathcal{P}} (\mathbf{x}\times\mathbf{t}+\mathbf{m})\, \mathrm{d}a + \int_{\mathcal{P}} (\mathbf{x}\times\mathbf{b}+\mathbf{c})\, \mathrm{d}v = \mathbf{0}, \tag{6.25b}$$

where $\mathbf{m} = \mathbf{n}\times\mathbf{s}$ and $\mathbf{c} = \mathbf{n}\times\boldsymbol{\beta}$.

The surface traction $\mathbf{t}$ is related to the Cauchy stress $\boldsymbol{\sigma}$ as

$$\mathbf{t} = \boldsymbol{\sigma}\overline{\mathbf{n}}, \tag{6.26}$$

where $\overline{\mathbf{n}}$ is the normal direction of the surface.

Similarly, the surface director-traction $\mathbf{s}$ is related to the second-order tensor $\mathbf{M}$ as

$$\mathbf{m} = \mathbf{M}\overline{\mathbf{n}}. \tag{6.27}$$

Combining Eqs. (6.25a) and (6.26) gives

$$\mathrm{div}\boldsymbol{\sigma} + \mathbf{b} = \mathbf{0}. \tag{6.28}$$

Similarly, combining Eqs. (6.25b) and (6.27) yields

$$\mathrm{div}\mathbf{M} + \mathbf{c} + 2\boldsymbol{\omega}^{\sigma} = \mathbf{0}, \tag{6.29}$$

where $\boldsymbol{\omega}^{\sigma}$ is the axial vector of skw $\boldsymbol{\sigma}$. The balance of linear momentum has the same form as shown in Chapter 1, while the balance of angular momentum is not the same as that shown in Chapter 1 due to the existence of director rotation.

6.4.3 Balance of Energy and the Second Law of Thermodynamics

We then try to obtain the internal power. Starting from Cauchy's postulate $\boldsymbol{\sigma}$ as $\mathbf{t} = \boldsymbol{\sigma}\overline{\mathbf{n}}$, the external power can be reformulated by employing the divergence theorem as

$$\int_{\partial\mathcal{P}} \mathbf{t}\cdot\mathbf{v}\, \mathrm{d}a = \int_{\mathcal{P}} \mathrm{div}(\boldsymbol{\sigma}^{\mathrm{T}}\mathbf{v})\, \mathrm{d}v = \int_{\mathcal{P}} (\boldsymbol{\sigma} : \nabla\mathbf{v} + \mathbf{v}\cdot \mathrm{div}\, \boldsymbol{\sigma})\, \mathrm{d}v. \tag{6.30}$$

For further analysis, it is beneficial to express the director rate $\dot{\mathbf{n}}$ in terms of the rotation tensor $\mathbf{R}$ and its material derivative $\dot{\mathbf{R}}$

$$\dot{\mathbf{n}} = \dot{\mathbf{R}}\mathbf{n}^0 = \dot{\mathbf{R}}\mathbf{R}^{\mathrm{T}}\mathbf{n} = \boldsymbol{\Omega}^{\mathrm{R}}\mathbf{n} = \boldsymbol{\omega}^{\mathrm{R}}\times\mathbf{n}. \tag{6.31}$$

Here, $\boldsymbol{\omega}^{\mathrm{R}}$ is the axial vector corresponding to the skew-symmetric angular velocity tensor $\boldsymbol{\Omega}^{\mathrm{R}} = \dot{\mathbf{R}}\mathbf{R}^{\mathrm{T}}$. Consequently, the contributions to the external power associated with the surface director force $\mathbf{s}$ and the body director force $\boldsymbol{\beta}$ can be expressed as

$$\int_{\partial\mathcal{P}} \mathbf{s}\cdot\dot{\mathbf{n}}\, \mathrm{d}a = \int_{\partial\mathcal{P}} \boldsymbol{\omega}^{\mathrm{R}}\cdot(\mathbf{n}\times\mathbf{s})\, \mathrm{d}a = \int_{\partial\mathcal{P}} \boldsymbol{\omega}^{\mathrm{R}}\cdot\mathbf{m}\, \mathrm{d}a, \tag{6.32a}$$

$$\int_{\mathcal{P}} \boldsymbol{\beta} \cdot \dot{\mathbf{n}} \, \mathrm{d}v = \int_{\mathcal{P}} \boldsymbol{\omega}^{\mathrm{R}} \cdot (\mathbf{n} \times \boldsymbol{\beta}) \, \mathrm{d}v = \int_{\mathcal{P}} \boldsymbol{\omega}^{\mathrm{R}} \cdot \mathbf{c} \, \mathrm{d}v. \tag{6.32b}$$

Utilizing the relationship $\mathbf{m} = \mathbf{M}\overline{\mathbf{n}}$, where $\mathbf{M}$ is a tensor related to the director stress, and applying the divergence theorem to the integral involving $\mathbf{s}$, we further obtain

$$\int_{\partial\mathcal{P}} \mathbf{s} \cdot \dot{\mathbf{n}} \, \mathrm{d}a = \int_{\partial\mathcal{P}} \boldsymbol{\omega}^{\mathrm{R}} \cdot \mathbf{m} \, \mathrm{d}a = \int_{\mathcal{P}} \mathrm{div}(\mathbf{M}^{\mathrm{T}}\boldsymbol{\omega}^{\mathrm{R}}) \, \mathrm{d}v = \int_{\mathcal{P}} (\mathbf{M} : \nabla\boldsymbol{\omega}^{\mathrm{R}} + \boldsymbol{\omega}^{\mathrm{R}} \cdot \mathrm{div}\mathbf{M}) \, \mathrm{d}v. \tag{6.33}$$

By substituting these derived expressions into the overall power balance equation and applying the local forms of the balance of linear momentum and the balance of angular momentum, the desired expression for the internal power can be derived as

$$P_{\mathrm{int}}(\mathbf{v}, \dot{\mathbf{n}}) = \int_{\mathcal{P}} (\boldsymbol{\sigma} : \nabla\mathbf{v} + \mathbf{M} : \nabla\boldsymbol{\omega}^{\mathrm{R}} - 2\,\boldsymbol{\omega}^{\sigma} \cdot \boldsymbol{\omega}^{\mathrm{R}}) \, \mathrm{d}v. \tag{6.34}$$

Under the assumptions of quasistatic and isothermal conditions, the Clausius–Duhem inequality imposes that the difference between the internal power and the rate of change of the Helmholtz free energy must be nonnegative. In its local form, this gives rise to the reduced dissipation inequality:

$$D_{\mathrm{int}} = -\frac{1}{J}\dot{\Psi} + \boldsymbol{\sigma} : \nabla\mathbf{v} + \mathbf{M} : \nabla\boldsymbol{\omega}^{\mathrm{R}} - 2\,\boldsymbol{\omega}^{\sigma} \cdot \boldsymbol{\omega}^{\mathrm{R}} \geq 0. \tag{6.35}$$

6.4.4 Constitutive Equations

We assume that the free energy per unit reference volume, Ψ, is a function of the total deformation gradient $\mathbf{F}$, the elastic part $\mathbf{F}^{\mathrm{e}}$, the director vector $\mathbf{n}$, its gradient $\nabla\mathbf{n}$, and an initial director $\mathbf{n}^0$ as a fixed parameter

$$\Psi = \hat{\Psi}(\mathbf{F}, \mathbf{F}^{\mathrm{e}}, \mathbf{n}, \nabla\mathbf{n}; \mathbf{n}^0). \tag{6.36}$$

It is important to note that the dependence of the free energy density on the internal variable $\mathbf{F}^{\mathrm{v}}$ arises through its influence on $\mathbf{F}^{\mathrm{e}}$ and the multiplicative decomposition (Eq. 6.21).

The rate of change of the free energy $\dot{\Psi}$ can be expressed as

$$\dot{\Psi} = \frac{\partial\Psi}{\partial\mathbf{F}} : \dot{\mathbf{F}} + \frac{\partial\Psi}{\partial\mathbf{F}^{\mathrm{e}}} : \dot{\mathbf{F}}^{\mathrm{e}} + \frac{\partial\Psi}{\partial\mathbf{n}} \cdot \dot{\mathbf{n}} + \frac{\partial\Psi}{\partial\nabla\mathbf{n}} : \dot{\overline{\nabla\mathbf{n}}}. \tag{6.37}$$

The expansion of the preceding equation, involving the expansion of the rate of change of $\mathbf{F}^{\mathrm{e}}$, expressing the rates of change of the director and its gradient in terms of the rate of change of rotation, and grouping terms leads to

$$\begin{aligned}\dot{\Psi} = {} & \left(\frac{\partial\Psi}{\partial F_{iA}} + \frac{\partial\Psi}{\partial F^{\mathrm{e}}_{i\alpha}} F^{v-1}_{A\alpha} - \frac{\partial\Psi}{\partial\mathbf{n}_{k,j}} \mathbf{n}_{k,i} F^{-1}_{Aj}\right) \dot{F}_{iA} + \left(\frac{\partial\Psi}{\partial\mathbf{n}_i}\mathbf{n}_j + \frac{\partial\Psi}{\partial\mathbf{n}_{i,k}}\mathbf{n}_{j,k}\right)\Omega^R_{ij} \\ & + \frac{\partial\Psi}{\partial\mathbf{n}_{i,k}}\mathbf{n}_j\Omega^R_{ij,k} - \frac{\partial\Psi}{\partial F^{\mathrm{e}}_{i\alpha}} F^{\mathrm{e}}_{i\beta} L^{\mathrm{v}}_{\beta\alpha}.\end{aligned} \tag{6.38}$$

Combining Eqs. (6.35) and (6.38) yields

$$\left[\sigma_{ij}F^{-1}_{Aj}-\frac{1}{J}\left(\frac{\partial\Psi}{\partial F_{iA}}+\frac{\partial\Psi}{\partial F^{e}_{i\alpha}}F^{v-1}_{A\alpha}-\frac{\partial\Psi}{\partial\mathbf{n}_{k,j}}\mathbf{n}_{k,j}F^{-1}_{Aj}\right)\right]\dot{F}_{iA}$$
$$-\left[\sigma_{[ij]}+\frac{1}{J}\left(\frac{\partial\Psi}{\partial\mathbf{n}_{i}}\mathbf{n}_{j}+\frac{\partial\Psi}{\partial\mathbf{n}_{i,k}}\mathbf{n}_{j,k}\right)\right]\Omega^{R}_{ij} \tag{6.39}$$
$$-\left[\frac{1}{2}e_{lij}M_{lk}+\frac{1}{J}\frac{\partial\Psi}{\partial\mathbf{n}_{i,k}}\mathbf{n}_{j}\right]\Omega^{R}_{ij,k}+\frac{1}{J}\frac{\partial\Psi}{\partial F^{e}_{i\alpha}}F^{e}_{i\beta}L^{v}_{\beta\alpha}\geq 0,$$

where $\sigma_{[ij]}=(\text{skw}\,\boldsymbol{\sigma})_{ij}$.

The nonnegative dissipation requirement (Eq. 6.39) can be satisfied by ensuring that the terms multiplying the independent rates are nonnegative separately, resulting in the following conditions:

$$\left[\sigma_{ij}F^{-1}_{Aj}-\frac{1}{J}\left(\frac{\partial\Psi}{\partial F_{iA}}+\frac{\partial\Psi}{\partial F^{e}_{i\alpha}}F^{v-1}_{A\alpha}-\frac{\partial\Psi}{\partial\mathbf{n}_{k,j}}\mathbf{n}_{k,j}F^{-1}_{Aj}\right)\right]\dot{F}_{iA}$$
$$-\left[\sigma_{[ij]}+\frac{1}{J}\left(\frac{\partial\Psi}{\partial\mathbf{n}_{i}}\mathbf{n}_{j}+\frac{\partial\Psi}{\partial\mathbf{n}_{i,k}}\mathbf{n}_{j,k}\right)\right]\Omega^{R}_{ij} \tag{6.40}$$
$$-\left[\frac{1}{2}e_{lij}M_{lk}+\frac{1}{J}\frac{\partial\Psi}{\partial\mathbf{n}_{i,k}}\mathbf{n}_{j}\right]\Omega^{R}_{ij,k}\geq 0,$$

and independently

$$\frac{\partial\Psi}{\partial F^{e}_{i\alpha}}F^{e}_{i\beta}L^{v}_{\beta\alpha}\geq 0. \tag{6.41}$$

These two inequalities (Eqs. 6.40 and 6.41) together provide a sufficient set of conditions to satisfy the overall dissipation inequality (Eq. 6.39).

Let us now introduce the constitutive response functions for the Cauchy stress $\boldsymbol{\sigma}$ and the couple stress tensor $\mathbf{M}$ as functions of the kinematic variables and their objective rates

$$\boldsymbol{\sigma}=\hat{\boldsymbol{\sigma}}(\mathbf{F},\mathbf{F}^{e},\mathbf{n},\nabla\mathbf{n},\overset{\circ}{\mathbf{n}},\overset{\circ}{\overline{\nabla\mathbf{n}}}), \tag{6.42}$$

$$\mathbf{M}=\hat{\mathbf{M}}(\mathbf{F},\mathbf{F}^{e},\mathbf{n},\nabla\mathbf{n},\overset{\circ}{\mathbf{n}},\overset{\circ}{\overline{\nabla\mathbf{n}}}), \tag{6.43}$$

where $\overset{\circ}{\overline{\mathbf{n}}}$ and $\overset{\circ}{\overline{\nabla\mathbf{n}}}$ are (at the moment arbitrary) objective rates for $\mathbf{n}$ and $\nabla\mathbf{n}$, respectively.

We can further decompose these stress tensors into equilibrium $(\boldsymbol{\sigma}^{\circ},\mathbf{M}^{\circ})$ and viscous $(\boldsymbol{\sigma}^{v},\mathbf{M}^{v})$ parts:

$$\boldsymbol{\sigma}=\boldsymbol{\sigma}^{\circ}+\boldsymbol{\sigma}^{v}, \tag{6.44a}$$

$$\mathbf{M}=\mathbf{M}^{\circ}+\mathbf{M}^{v}. \tag{6.44b}$$

The equilibrium part depends only on the current state variables:

$$\boldsymbol{\sigma}^{\circ}=\hat{\boldsymbol{\sigma}}^{\circ}(\mathbf{F},\mathbf{F}^{e},\mathbf{n},\nabla\mathbf{n})=\hat{\boldsymbol{\sigma}}(\mathbf{F},\mathbf{F}^{e},\mathbf{n},\nabla\mathbf{n},\mathbf{0},\mathbf{0}), \tag{6.45a}$$

$$\mathbf{M}^{\circ}=\hat{\mathbf{M}}^{\circ}(\mathbf{F},\mathbf{F}^{e},\mathbf{n},\nabla\mathbf{n})=\hat{\mathbf{M}}(\mathbf{F},\mathbf{F}^{e},\mathbf{n},\nabla\mathbf{n},\mathbf{0},\mathbf{0}). \tag{6.45b}$$

The viscous part captures the rate-dependent behavior:

$$\boldsymbol{\sigma}^{\mathrm{v}} = \hat{\boldsymbol{\sigma}}^{\mathrm{v}}(\mathbf{F}, \mathbf{F}^{\mathrm{e}}, \mathbf{n}, \nabla\mathbf{n}, \overset{\circ}{\mathbf{n}}, \overset{\circ}{\overline{\nabla\mathbf{n}}}) = \hat{\boldsymbol{\sigma}}(\mathbf{F}, \mathbf{F}^{\mathrm{e}}, \mathbf{n}, \nabla\mathbf{n}, \overset{\circ}{\mathbf{n}}, \overset{\circ}{\overline{\nabla\mathbf{n}}}) - \hat{\boldsymbol{\sigma}}^{\circ}(\mathbf{F}, \mathbf{F}^{\mathrm{e}}, \mathbf{n}, \nabla\mathbf{n}), \tag{6.46a}$$

$$\mathbf{M}^{\mathrm{v}} = \hat{\mathbf{M}}^{\mathrm{v}}(\mathbf{F}, \mathbf{F}^{\mathrm{e}}, \mathbf{n}, \nabla\mathbf{n}, \overset{\circ}{\mathbf{n}}, \overset{\circ}{\overline{\nabla\mathbf{n}}}) = \hat{\mathbf{M}}(\mathbf{F}, \mathbf{F}^{\mathrm{e}}, \mathbf{n}, \nabla\mathbf{n}, \overset{\circ}{\mathbf{n}}, \overset{\circ}{\overline{\nabla\mathbf{n}}}) - \hat{\mathbf{M}}^{\circ}(\mathbf{F}, \mathbf{F}^{\mathrm{e}}, \mathbf{n}, \nabla\mathbf{n}). \tag{6.46b}$$

Given that $\dot{\mathrm{F}}_{iA}$, Ω^{R}_{ij}, and $\Omega^{\mathrm{R}}_{ij,k}$ are arbitrary and mutually independent, and considering the skew-symmetric nature of Ω^{R}_{ij} and $\Omega^{\mathrm{R}}_{ij,k}$ in ij, we can apply the Coleman–Noll argument to inequality (Eq. 6.40). This application demonstrates the necessity of the following relationships:

$$\boldsymbol{\sigma}^{\circ} = \frac{1}{J}\left(\frac{\partial\Psi}{\partial\mathbf{F}}\mathbf{F}^{\mathrm{T}} + \frac{\partial\Psi}{\partial\mathbf{F}^{\mathrm{e}}}\mathbf{F}^{\mathrm{eT}} - (\nabla\mathbf{n})^{\mathrm{T}}\frac{\partial\Psi}{\partial\nabla\mathbf{n}}\right), \tag{6.47}$$

$$\mathrm{skw}\,\boldsymbol{\sigma}^{\circ} = -\frac{1}{J}\,\mathrm{skw}\left(\frac{\partial\Psi}{\partial\mathbf{n}}\otimes\mathbf{n} + \frac{\partial\Psi}{\partial\nabla\mathbf{n}}(\nabla\mathbf{n})^{\mathrm{T}}\right), \tag{6.48}$$

$$\mathbf{M}^{\circ} = \frac{1}{J}\mathbf{n}\times\frac{\partial\Psi}{\partial\nabla\mathbf{n}}. \tag{6.49}$$

After applying the Coleman–Noll argument, the remaining inequality takes the form

$$\sigma^{\mathrm{v}}_{(ij)}D_{ij} + \sigma^{\mathrm{v}}_{[ij]}(W_{ij} - \Omega^{\mathrm{R}}_{ij}) - \frac{1}{2}e_{lij}M^{\mathrm{v}}_{lk}\Omega^{\mathrm{R}}_{ij,k} \geq 0, \tag{6.50}$$

where $\mathbf{D} = \mathrm{sym}(\nabla\mathbf{v})$, $\mathbf{W} = \mathrm{skw}(\nabla\mathbf{v})$, and $\sigma^{\mathrm{v}}_{(ij)} = (\mathrm{sym}\,\boldsymbol{\sigma}^{\mathrm{v}})_{ij}$.

Furthermore, we posit that the objective rates $\overset{\circ}{\mathbf{n}}$ and $\overset{\circ}{\overline{\nabla\mathbf{n}}}$ are independent of $\mathbf{D}$. This assumption is equivalent to stating that the dissipation arising from network deformation and director rotation are decoupled. A sequential application of the Coleman–Noll argument, with arbitrary choices for D_{ij}, W_{ij}, Ω^{R}_{ij}, and $\Omega^{\mathrm{R}}_{ij,k}$, yields

$$\mathrm{sym}\,\boldsymbol{\sigma}^{\mathrm{v}} = \mathbf{0}. \tag{6.51}$$

Consequently, the inequality that remains to be satisfied is

$$\sigma^{\mathrm{v}}_{[ij]}(W_{ij} - \Omega^{\mathrm{R}}_{ij}) - \frac{1}{2}e_{lij}M^{\mathrm{v}}_{lk}\Omega^{\mathrm{R}}_{ij,k} \geq 0. \tag{6.52}$$

The inequality is completed together with the internal variable dissipation inequality (Eq. 6.41).

6.4.5 Simplification for the Non-gradient Case

Gradients play a significant role in LCEs, particularly in resolving the degeneracy of equilibrium behavior (as demonstrated, for instance, by the neoclassical model). These gradient effects can manifest as purely elastic, purely viscous, or a combination of both, necessitating a constitutive law for $\mathbf{M}^{\mathrm{v}}$ in the latter case. In the absence of definitive experimental evidence, the attention is restricted to the non-gradient scenario. Therefore, the models are confined to those where $\boldsymbol{\sigma}^{\mathrm{v}}$ and $\mathbf{M}^{\mathrm{v}}$ are independent of $\nabla\mathbf{n}$. Under this constraint, $\mathbf{M}^{\mathrm{v}}$ becomes independent of the gradient of $\boldsymbol{\Omega}^{\mathrm{R}}$, and we can directly infer that $\mathbf{M}^{\mathrm{v}} = \mathbf{0}$ through a further application of the Coleman–Noll argument to the final term in Eq. (6.52). This simplification reduces Eq. (6.52) to the requirement that

$$\operatorname{skw}\boldsymbol{\sigma}^{\mathrm{v}} : (\mathbf{W} - \boldsymbol{\Omega}^{\mathrm{R}}) \geq 0. \tag{6.53}$$

To fulfill Eq. (6.53), we select a suitable objective rate that is independent of $\mathbf{D}$, specifically $\mathring{\mathbf{n}} = \dot{\mathbf{n}} - \mathbf{W}\mathbf{n}$. This choice leads to the following constitutive relation:

$$\operatorname{skw}\boldsymbol{\sigma}^{\mathrm{v}} = \eta_{\mathrm{D}} \operatorname{skw}(\mathbf{n} \otimes \mathring{\mathbf{n}}), \tag{6.54}$$

where η_{D} represents the director viscosity.

Evaluating the left-hand side of Eq. (6.53), we obtain

$$\begin{aligned}
\operatorname{skw}\boldsymbol{\sigma}^{\mathrm{v}} : (\mathbf{W} - \boldsymbol{\Omega}^{\mathrm{R}}) &= \eta_{\mathrm{D}} \operatorname{skw}(\mathbf{n} \otimes \mathring{\mathbf{n}}) : (\mathbf{W} - \boldsymbol{\Omega}^{\mathrm{R}}) \\
&= \eta_{\mathrm{D}} (\mathbf{n} \otimes \mathring{\mathbf{n}}) : (\mathbf{W} - \boldsymbol{\Omega}^{\mathrm{R}}) \\
&= \eta_{\mathrm{D}} \mathring{\mathbf{n}} \cdot (\boldsymbol{\Omega}^{\mathrm{R}} - \mathbf{W})\mathbf{n} \\
&= \eta_{\mathrm{D}} \mathring{\mathbf{n}} \cdot (\dot{\mathbf{n}} - \mathbf{W}\mathbf{n}) \\
&= \eta_{\mathrm{D}} \mathring{\mathbf{n}} \cdot \mathring{\mathbf{n}}.
\end{aligned} \tag{6.55}$$

Given that $\mathbf{W}$ and $\boldsymbol{\Omega}^{\mathrm{R}}$ are skew-symmetric and that $\dot{\mathbf{n}} = \boldsymbol{\Omega}^{\mathrm{R}}\mathbf{n}$, Eq. (6.53) simplifies to the requirement

$$\eta_{\mathrm{D}} \, ||\mathbf{n}||^2 \geq 0. \tag{6.56}$$

This condition is inherently satisfied, provided that the director viscosity η_{D} is nonnegative.

To fulfill inequality in Eq. (6.41), we propose the following evolution equation for the polymeric network:

$$\mathbf{L}^{\mathrm{v}} = \frac{1}{\eta_{\mathrm{N}}} \mathbf{F}^{\mathrm{eT}} \frac{\partial \Psi}{\partial \mathbf{F}^{\mathrm{e}}}, \tag{6.57}$$

where $\eta_{\mathrm{N}} > 0$ represents the network viscosity.

Consequently, the total dissipation within the model is expressed as

$$D_{\mathrm{int}} = \eta_{\mathrm{N}} \, ||\mathbf{L}^{\mathrm{v}}||^2 / J + \eta_{\mathrm{D}} \, ||\mathring{\mathbf{n}}||^2. \tag{6.58}$$

In this expression, the first term signifies the dissipation due to the network, while the second term corresponds to the director dissipation.

It is further assumed that the free energy density Ψ does not explicitly depend on $\nabla\mathbf{n}$, and also that body director forces are absent.

The primary consequence of the first assumption is the vanishing of the couple stress tensor $\mathbf{M}$. Utilizing this result in conjunction with the second assumption and the balance of angular momentum, it can be further deduced that, under these particular conditions, the Cauchy stress tensor $\boldsymbol{\sigma}$ becomes symmetric.

Employing Eqs. (6.47), (6.51), and (6.54), the resulting expression for the stress tensor is

$$\boldsymbol{\sigma}=\frac{1}{J}\left(\frac{\partial\Psi}{\partial\mathbf{F}}\mathbf{F}^{\mathrm{T}}+\frac{\partial\Psi}{\partial\mathbf{F}^{\mathrm{e}}}\mathbf{F}^{\mathrm{eT}}\right)+\eta_D\,\mathrm{skw}(\mathbf{n}\otimes\overset{\circ}{\mathbf{n}})=\frac{1}{J}\,\mathrm{sym}\left(\frac{\partial\Psi}{\partial\mathbf{F}}\mathbf{F}^{\mathrm{T}}+\frac{\partial\Psi}{\partial\mathbf{F}^{\mathrm{e}}}\mathbf{F}^{\mathrm{eT}}\right). \tag{6.59}$$

The symmetry of the stress tensor can also be employed to derive an explicit evolution law for the director, using the condition that $\mathrm{skw}\,\boldsymbol{\sigma}=\mathrm{skw}\,\boldsymbol{\sigma}^{\circ}+\mathrm{skw}\,\boldsymbol{\sigma}^{\mathrm{v}}=\mathbf{0}$.

Substituting Eqs. (6.48) and (6.54) into this condition and applying the resulting tensor equation to the director vector $\mathbf{n}$, we obtain

$$\overset{\circ}{\mathbf{n}}=-\frac{1}{J\eta_D}\left[\frac{\partial\Psi}{\partial\mathbf{n}}-\left(\frac{\partial\Psi}{\partial\mathbf{n}}\cdot\mathbf{n}\right)\mathbf{n}\right]. \tag{6.60}$$

In summary, the constitutive equations for the gradient-independent case consist of the stress expression (Eq. 6.59), the viscoelastic flow rule for the polymer network (Eq. 6.57), and the viscous flow relation for the liquid crystal director (Eq. 6.60).

6.4.6 Free Energy Density

The network deformation gradient $\mathbf{F}_N$ can also be represented by the elastic part and viscous part, which is

$$\mathbf{F}_N=\mathbf{F}_N^{\mathrm{e}}\mathbf{F}_N^{\mathrm{v}}. \tag{6.61}$$

Assuming the reference director $\mathbf{n}_0$ is unaffected by viscous deformation $\mathbf{F}^{\mathrm{v}}$, the viscous network deformation tensor $\mathbf{F}_N^{\mathrm{v}}$ can be defined as

$$\mathbf{F}_N^{\mathrm{v}}=\boldsymbol{\ell}_0^{-1/2}\,\mathbf{F}^{\mathrm{v}}\,\boldsymbol{\ell}_0^{1/2}. \tag{6.62}$$

And the elastic network deformation tensor is

$$\mathbf{F}_N^{\mathrm{e}}=\boldsymbol{\ell}^{-1/2}\,\mathbf{F}^{\mathrm{e}}\,\boldsymbol{\ell}_0^{1/2}. \tag{6.63}$$

Then the general form of the free energy density can be divided into the equilibrium part and the nonequilibrium part as

$$\Psi=\Psi^{\mathrm{eq}}(\mathbf{C},\mathbf{C}_N)+\Psi^{\mathrm{neq}}(\mathbf{C}^{\mathrm{e}},\mathbf{C}_N^{\mathrm{e}}), \tag{6.64}$$

where $\mathbf{C}^{\mathrm{e}}=\mathbf{F}^{\mathrm{eT}}\mathbf{F}^{\mathrm{e}}$ and $\mathbf{C}_N^{\mathrm{e}}=\mathbf{F}_N^{\mathrm{eT}}\mathbf{F}_N^{\mathrm{e}}$.

To describe the strain-stiffening response at large deformation caused by chain alignment, the modified Gent model is applied to the equilibrium and nonequilibrium free energy density:

$$\Psi^{\mathrm{eq}}(\mathbf{C},\mathbf{C}_N)=-\frac{G^{\mathrm{eq}}}{2}I_m\ln\left(1-\frac{I_{N1}-3}{I_m}\right)-p(J-1), \tag{6.65}$$

$$\Psi^{\mathrm{neq}}(\mathbf{C}^{\mathrm{e}},\mathbf{C}_N^{\mathrm{e}})=-\frac{G^{\mathrm{neq}}}{2}I_m\ln\left(1-\frac{I_{N1}^{\mathrm{e}}-3}{I_m}\right)-G^{\mathrm{neq}}\ln J^{\mathrm{e}}, \tag{6.66}$$

where p is the Lagrange multiplier used to enforce the incompressibility constraint $J=\det\mathbf{F}=1$, $J^{\mathrm{e}}=\det\mathbf{F}^{\mathrm{e}}$, G^{eq} and G^{neq} are the equilibrium and nonequilibrium shear moduli, respectively, I_m is the stiffening parameter in the Gent model, and $I_{N1}^{\mathrm{e}}=\mathrm{Tr}\left(\mathbf{C}_N^{\mathrm{e}}\right)$.

Inserting Eqs. (6.65) and (6.66) into Eq. (6.59), the Cauchy stress can be represented as

$$\boldsymbol{\sigma} = \boldsymbol{\sigma}^{\text{eq}} + \boldsymbol{\sigma}^{\text{neq}}, \tag{6.67a}$$

$$\begin{aligned} \boldsymbol{\sigma}^{\text{eq}} &= \text{sym}(\frac{\partial \Psi}{\partial \mathbf{F}}\mathbf{F}^{\text{T}}) \\ &= \frac{G^{\text{eq}}}{1-Q}\frac{I_m}{I_m-(I_{N_1}-3)}\left[\mathbf{F}\ell_0\mathbf{F}^{\text{T}} - \frac{3Q}{1+2Q}\text{sym}(\mathbf{n}\otimes\mathbf{F}\ell_0\mathbf{F}^{\text{T}}\mathbf{n})\right] - p\mathbf{I}, \end{aligned} \tag{6.67b}$$

$$\begin{aligned} \boldsymbol{\sigma}^{\text{neq}} &= \text{sym}(\frac{\partial \Psi}{\partial \mathbf{F}^{\text{e}}}\mathbf{F}^{\text{e}^{\text{T}}}) \\ &= \frac{G^{\text{neq}}}{1-Q}\frac{I_m}{I_m-(I^{\text{e}}_{N1}-3)}\left[\mathbf{F}^{\text{e}}\ell_0\mathbf{F}^{\text{e}^{\text{T}}} - \frac{3Q}{1+2Q}\text{sym}(\mathbf{n}\otimes\mathbf{F}^{\text{e}}\ell_0\mathbf{F}^{\text{e}^{\text{T}}}\mathbf{n})\right] - G^{\text{neq}}\mathbf{I}. \end{aligned} \tag{6.67c}$$

The network flow rule is represented as

$$\mathbf{L}^{\text{v}} = \frac{G^{\text{neq}}}{\eta_N}\left(\frac{I_m}{I_m-(I^{\text{e}}_{N1}-3)}\mathbf{F}^{\text{e}^{\text{T}}}\ell^{-1}\mathbf{F}^{\text{e}}\ell_0 - \mathbf{I}\right). \tag{6.68}$$

Substituting Eqs. (6.65) and (6.66) into Eq. (6.60), the following explicit form is obtained for director rotation as

$$\dot{\mathbf{n}} = \mathbf{W}\mathbf{n} - \frac{1}{J\eta_D}\underbrace{[\frac{\partial \Psi^{\text{eq}}}{\partial \mathbf{n}} - (\frac{\partial \Psi^{\text{eq}}}{\partial \mathbf{n}}\cdot\mathbf{n})\mathbf{n}]}_{\mathbf{m}^{\text{eq}}} - \frac{1}{J\eta_D}\underbrace{[\frac{\partial \Psi^{\text{neq}}}{\partial \mathbf{n}} - (\frac{\partial \Psi^{\text{eq}}}{\partial \mathbf{n}}\cdot\mathbf{n})\mathbf{n}]}_{\mathbf{m}^{\text{neq}}}, \tag{6.69}$$

$$\mathbf{m}^{\text{eq}} = -G^{\text{eq}}\frac{3Q}{(1-Q)(1+2Q)}\frac{I_m}{I_m-(I_{N1}-3)}(\mathbf{F}\cdot\mathbf{n} - (\mathbf{n}\cdot\mathbf{F}\cdot\mathbf{n})\mathbf{n}), \tag{6.70}$$

$$\mathbf{m}^{\text{neq}} = -G^{\text{neq}}\frac{3Q}{(1-Q)(1+2Q)}\frac{I_m}{I_m-(I^{\text{e}}_{N1}-3)}(\mathbf{F}^{\text{e}}\cdot\mathbf{n} - (\mathbf{n}\cdot\mathbf{F}^{\text{e}}\cdot\mathbf{n})\mathbf{n}). \tag{6.71}$$

Considering the power-law rate-dependence of the peak stress, the director viscosity is represented as

$$\eta_D = \eta_{D0}\exp\left(-\frac{\sqrt{\boldsymbol{\sigma}:\boldsymbol{\sigma}}}{\text{k}_S}\right). \tag{6.72}$$

The above model combines the time-dependent dissipation of director rotation with the viscoelasticity of the polymer network, providing a more accurate and physically relevant description of the mechanical response of LCEs. The model is calibrated with the experimental results of a type of monodomain LCEs. The obtained model parameters are set as $G^{\text{eq}} = 0.25$ MPa, $G^{\text{neq}} = 1.25$ MPa, $I_m = 5.7$, $\eta_N = 800$ MPa · s, Q=0.62, $\eta_{D0} = 40$ MPa · s, $k_S = 0.7$ MPa. The predicted results of the above model are shown in Figure 6.6 for materials with the director initially perpendicular to applied loading, which captures the main features in experiments. Based on this framework, Kang et al. [38] further incorporated the thermal-order coupling effect into the model for monodomain LCEs, which is not further elaborated in this chapter.

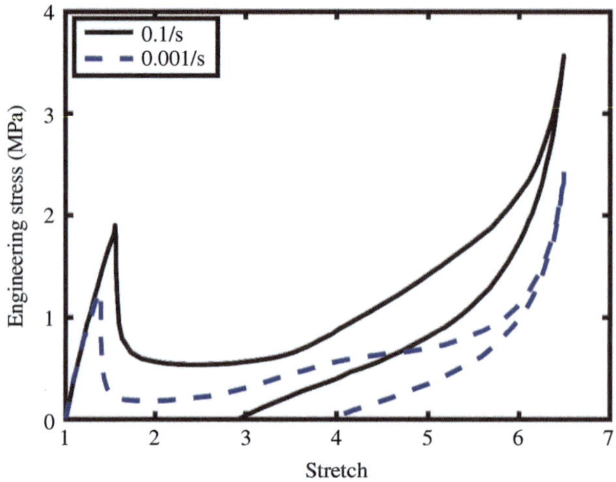

Figure 6.6 The prediction results of uniaxial tension of LCE with director initially perpendicular to applied loading. *Source*: Wang et al. (2022)/with permission of Elsevier.

6.5 Theory for Polydomain Liquid Crystal Elastomers

To understand the origins of polydomain LCEs, consider a small representative volume, as illustrated in Figure 6.7. At high temperatures, the liquid crystal molecules are disordered. If cross-linking occurs in this state, the elastomer remains isotropic. Upon cooling, the liquid crystals align into a nematic phase, establishing a uniform director field within each domain. However, because the high-temperature cross-linked state is isotropic, the nematic director can orient in arbitrary directions. Consequently, multiple single domains with different director orientations coexist, resulting in polydomain LCEs. Additionally, polydomain LCEs can also be formed by cross-linking in the nematic state. The macroscopic properties of LCEs are influenced by both the cross-linking conditions and the resulting polydomain structure.

Based on the neoclassical model, Biggins et al. [30] extended the energy formulation from monodomain LCEs to polydomain LCEs. The cross-linked state is defined as the elastic reference state. The deformation gradient from this state is further denoted as $\boldsymbol{\gamma}$. For a polydomain LCE cross-linked in the isotropic state, the energy function takes the form

$$\Psi = \frac{1}{2} G \,\mathrm{Tr}\left(\boldsymbol{\gamma}^T \cdot \boldsymbol{\ell}^{-1} \cdot \boldsymbol{\gamma}\right), \tag{6.73}$$

which is identical to the corresponding energy function for monodomain LCEs. This lack of intrinsic spatial variation in the energy is expected, given that the cross-linked state is fully isotropic and homogeneous.

For a polydomain LCE cross-linked in the nematic state, the energy function is modified to account for the nematic director field and is expressed as

$$\Psi = \frac{1}{2} G \,\mathrm{Tr}\left(\boldsymbol{\gamma} \cdot \boldsymbol{\ell}_0 \cdot \boldsymbol{\gamma}^T \cdot \boldsymbol{\ell}^{-1}\right), \tag{6.74}$$

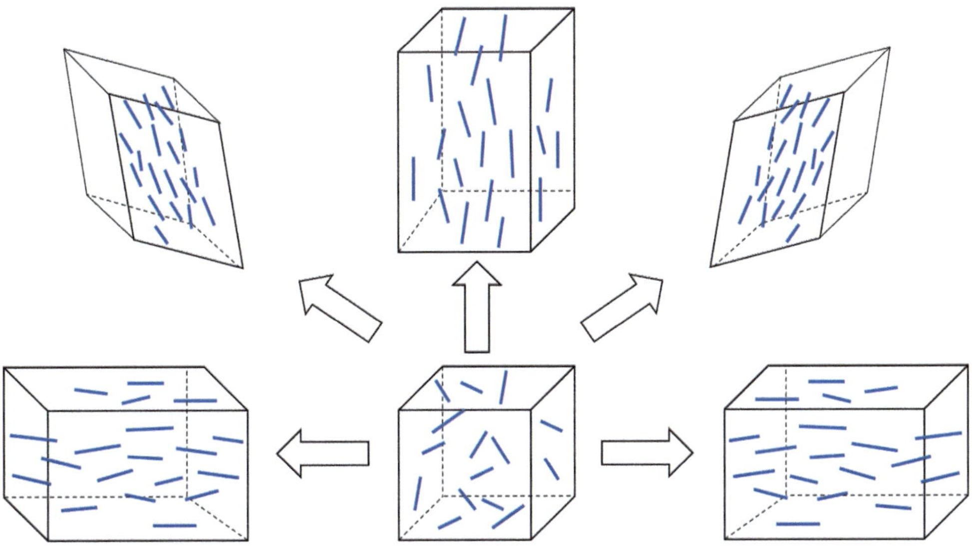

Figure 6.7 Different domain orientations within various regions of polydomain liquid crystal elastomers.

where ℓ_0 is derived from $\hat{\mathbf{n}}_0$, the nematic director at cross-linking. Unlike the energy for monodomain LCEs, this energy function exhibits a strong dependence on the director orientations within each domain. This distinction arises because, in polydomain LCEs, the directors in different domains are different, whereas in monodomain LCEs, there is a single, consistent director orientation throughout the material.

Real polymer chains are neither perfectly Gaussian nor compositionally homogeneous. In nematic elastomers, this results in a preferred director orientation in the nematic state, known as nonideality. For elastomers cross-linked in the isotropic state, this preference is weak and spatially random. However, in elastomers cross-linked in the nematic state, the director at the time of cross-linking is remembered, and the free energy is minimized when the director aligns with this orientation. To account for this, a nonideal term is added to the free energy, favoring alignment along the initial director direction, $\hat{\mathbf{n}}_0$. The energy form of a polydomain LCE cross-linked in the nematic phase is then written as

$$\Psi = \frac{1}{2} G \,\mathrm{Tr}\left(\boldsymbol{\gamma} \cdot \ell_0 \cdot \boldsymbol{\gamma}^T \cdot \ell^{-1} + \alpha \gamma \cdot (\delta - \hat{\mathbf{n}}_0 \hat{\mathbf{n}}_0) \cdot \gamma^T \cdot \hat{\mathbf{n}}\hat{\mathbf{n}}\right). \tag{6.75}$$

The coefficient α determines the extent that the elastomer deviates from ideal soft behavior.

A similar energy form is given for isotropic genesis polydomain specimens:

$$\Psi = \frac{1}{2} G \,\mathrm{Tr}\left(\boldsymbol{\gamma}^T \cdot \ell^{-1} \cdot \boldsymbol{\gamma} + \alpha r^{1/3} \gamma \cdot (\delta - \hat{\mathbf{n}}_0 \hat{\mathbf{n}}_0) \cdot \gamma^T \cdot \hat{\mathbf{n}}\hat{\mathbf{n}}\right), \tag{6.76}$$

where r is the degree of anisotropy in the nematic polymer.

Thus, four distinct forms of energy can be derived, depending on the cross-linking state and the degree of nonideality:

$$\Psi = \begin{cases} \frac{1}{2} G \operatorname{Tr} (\gamma^T \cdot \ell^{-1} \cdot \gamma), & iI \\ \frac{1}{2} G \operatorname{Tr} (\gamma^T \cdot \ell^{-1} \cdot \gamma + \alpha r^{1/3} \gamma \cdot (\delta - \hat{\mathbf{n}}_0 \hat{\mathbf{n}}_0) \cdot \gamma^T \cdot \hat{\mathbf{n}}\hat{\mathbf{n}}), & nI \\ \frac{1}{2} G \operatorname{Tr} (\gamma \cdot \ell_0 \cdot \gamma^T \cdot \ell^{-1}), & iN \\ \frac{1}{2} G \operatorname{Tr} (\gamma \cdot \ell_0 \cdot \gamma^T \cdot \ell^{-1} + \alpha \gamma \cdot (\delta - \hat{\mathbf{n}}_0 \hat{\mathbf{n}}_0) \cdot \gamma^T \cdot \hat{\mathbf{n}}\hat{\mathbf{n}}). & nN \end{cases} \tag{6.77}$$

Here, iI/nI refers to ideal/nonideal elastomers cross-linked in the isotropic state, and iN/nN refers to ideal/nonideal elastomers cross-linked in the nematic state. In all cases, γ represents the deformation gradient relative to the natural cross-linked state. The elastomer is assumed to be incompressible; hence $\det(\gamma) = 1$. A factor of $r^{1/3}$ is factored out of the α term in the nI energy to ensure that, under full relaxation, an individual domain would align with the energy form of the nN case. This model, while providing valuable insights, falls short of capturing the full complexity of polydomain LCEs under realistic conditions. A major limitation arises from the fact that only the ideal isotropic origin (iI) system can be fully resolved, which restricts the applicability across a broader range of scenarios.

The rapid development of parallel computing, particularly through GPUs and specialized accelerators, has paved the way for innovative approaches to micromechanics problems. For instance, Zhou and Bhattacharya [31] proposed an operator splitting scheme that frames physical governing equations as incremental variational problems. Differential constraints, such as compatibility, are incorporated via an augmented generalized Lagrangian, and the resulting variational principle is solved using the multiplicative alternating direction method. This method offers novel perspectives on the microstructural evolution of LCEs.

Tokumoto et al. [39] applied biaxial tensile tests to characterize the mechanical response of polydomain LCEs. The results show that when polydomain LCEs are subjected to unequal biaxial stretch, there is a regime where the true stress in both principal directions remains equal, even when the stretches are unequal. Furthermore, the true stress depends only on the areal stretch (the product of the two stretches), independent of their individual values or loading history. Based on this, a corresponding theoretical model was developed. The energy is defined as follows:

$$\begin{aligned} W &= \int_V \Psi(\mathbf{F}, \mathbf{n}, \nabla \mathbf{n}, \mathbf{n}_0)\, \mathrm{d}V \\ &= \int_V \left(\frac{G}{2} \left(\operatorname{Tr}(\mathbf{F}^T \ell^{-1} \mathbf{F}) - 3\right) + \frac{\alpha}{2} \left(|\mathbf{F}^T \mathbf{n}|^2 - (\mathbf{n}_0 \cdot \mathbf{F}^T \mathbf{n})^2 \right) + \frac{K}{2} |\nabla \mathbf{n}|^2 \right) \mathrm{d}V, \end{aligned} \tag{6.78}$$

where $\mathbf{n}_0$ is a random unit vector field. The first term represents the Gaussian elasticity of the elastomer with shear modulus G. The second term, which depends on the parameter $\alpha > 0$, describes the random disorder of the network [40], leading to a preferred random director field $\mathbf{n}_0$. The final term represents the Frank elasticity, which penalizes changes in the director orientation under the single constant approximation.

In general deformation processes, this equation does not yield a direct solution. Therefore, Tokumoto et al. [39] employed backward Eulerian discretization to implicitly discretize these differential equations over time, rewriting them as variational problems to obtain the solution. The resulting predictions are shown in Figure 6.8, where the theoretically derived true stress–stretch relationship accurately describes the stress–stretch behavior observed in uniaxial experiments.

However, these microscopic approaches require substantial computational resources, posing challenges for their application in typical engineering contexts. Thus, it is still highly desirable to develop an explicit constitutive model for polydomain LCEs. Lee et al. [32] proposed a macroscopic constitutive model for polydomain LCEs. This model implicitly accounts for the complex evolution of domain modes and can accurately capture the complex response to multiaxial loading observed experimentally. To describe the deformation of individual domains, the anisotropy of polydomain LCEs is characterized by the right Cauchy–Green tensor **G**, which is defined at the mesoscale, as opposed to macroscopic deformation. **G** is a symmetric, positive-definite tensor that satisfies det(**G**) = 1. As discussed in Lee et al. [32], the spontaneous deformation state of each microscopic domain can be associated with the tensor **G**, leading to the following relationship:

$$\mathbf{G} = \mathbf{P}\mathbf{G}_0\mathbf{P}^{\mathrm{T}}, \tag{6.79}$$

where **P** represents a rotation matrix and $\mathbf{G}_0$ is a diagonal tensor with positive entries, defined in a fixed laboratory frame.

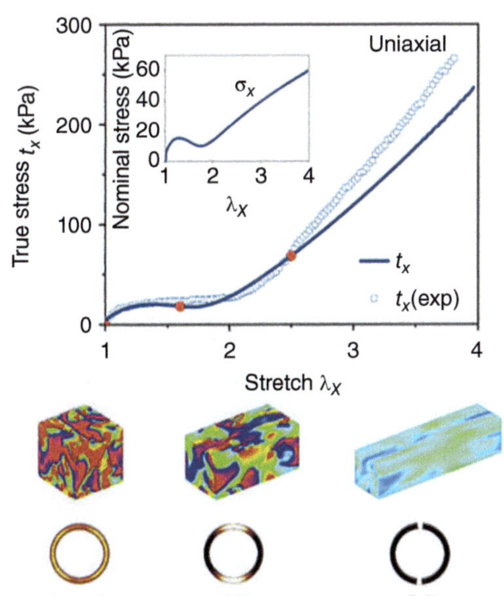

Figure 6.8 Comparison between the predicted and measured stress–stretch relations, and director distribution for various types of extension obtained by the full-field 3D simulations. *Source:* Tokumoto et al. (2021)/American Association for the Advancement of Science./CC BY 4.0.

Two state variables Λ and Δ are defined, corresponding to the maximum principal stretch and maximum principal area stretch within each domain, respectively. $\mathbf{G}_0$ can then be represented as

$$\mathbf{G}_0 = \Lambda^2 \mathbf{e}_1 \otimes \mathbf{e}_1 + \frac{\Delta^2}{\Lambda^2} \mathbf{e}_2 \otimes \mathbf{e}_2 + \frac{1}{\Delta^2} \mathbf{e}_3 \otimes \mathbf{e}_3, \tag{6.80}$$

where the ordering $\Lambda \geq \Delta/\Lambda \geq 1/\Delta$ holds. This imposes restrictions on the possible values of Λ and Δ. In the monodomain state, $\mathbf{G}$ reduces to ℓ. Incorporating the constraints $\Lambda \geq \Delta/\Lambda \geq 1/\Delta$, the allowable values for Λ and Δ are confined to the set $\mathcal{A}_G$:

$$(\Lambda, \Delta) \in \mathcal{A}_G = \{(\Lambda, \Delta) : \Delta \leq r^{1/6}, \Delta \leq \Lambda^2, \Delta \geq \sqrt{\Lambda}\}. \tag{6.81}$$

In contrast to other energy formulations, the model proposed by Lee et al. [32] establishes a direct connection between the mesoscale deformation $\mathbf{G}$ and the macroscopic deformation gradient $\mathbf{F}$, partitioning the total free energy into two distinct components:

$$W(\mathbf{F}, \Lambda, \Delta) = W_{\mathrm{e}}(\mathbf{F}, \Lambda, \Delta) + W_{\mathrm{r}}(\Lambda, \Delta). \tag{6.82}$$

The first term, W_{e}, represents the elastic energy stored in the polymer network due to deformation relative to the spontaneous deformation of the domains. The second term, W_{r}, accounts for the residual energy arising from imperfections and incompatibilities within the domains.

The free energy of the elastomer can then be expressed in the following form:

$$\begin{aligned} W_{\mathrm{e}}(\mathbf{F}, \Lambda, \Delta) &= \varphi\left(I_1(\mathbf{F}^{\mathrm{T}}\mathbf{G}\mathbf{F}), I_2(\mathbf{F}^{\mathrm{T}}\mathbf{G}\mathbf{F})\right) \\ &= \varphi\left(\frac{\lambda_1^2}{\Lambda^2} + \frac{\lambda_2^2\Lambda^2}{\Delta^2} + \lambda_3^2\Delta^2, \frac{\Lambda^2}{\lambda_1^2} + \frac{\Delta^2}{\lambda_2^2\Lambda^2} + \frac{1}{\lambda_3^2\Delta^2}\right), \end{aligned} \tag{6.83}$$

where φ is a nonnegative, convex, and monotonically increasing function of its arguments, with I_1 and I_2 representing the first two principal invariants.

The residual energy is expressed as

$$W_{\mathrm{r}} = C\frac{\Delta - 1}{(r^{1/6} - \Delta)^k}, \tag{6.84}$$

where $C > 0$ and $k > 1$ are material-specific constants to be determined. Experimental and numerical findings [31, 39] reveal that the stoichiometry of polydomain LCE cross-linking establishes the isotropic state ($\Delta = 1$) as the reference ground state.

The evolution of the state variables follows the relationship:

$$\nu_\Lambda \dot{\Lambda} = -\frac{\partial W}{\partial \Lambda}, \quad \nu_\Delta \dot{\Delta} = -\frac{\partial W}{\partial \Delta}, \tag{6.85}$$

where ν_Λ and ν_Δ are kinetic coefficients. In this manner, the complex evolution of the domain structure can be implicitly captured, allowing for an accurate representation of the observed response under multiaxial loading conditions, as illustrated in Figure 6.9. This approach ensures that the intricate interplay between the domains and external stresses is faithfully modeled, enabling predictions that align closely with experimental observations. We have also employed this model to simulate the inflation behaviors of polydomain LCE membranes, which shows full consistency between experiments and predictions [41].

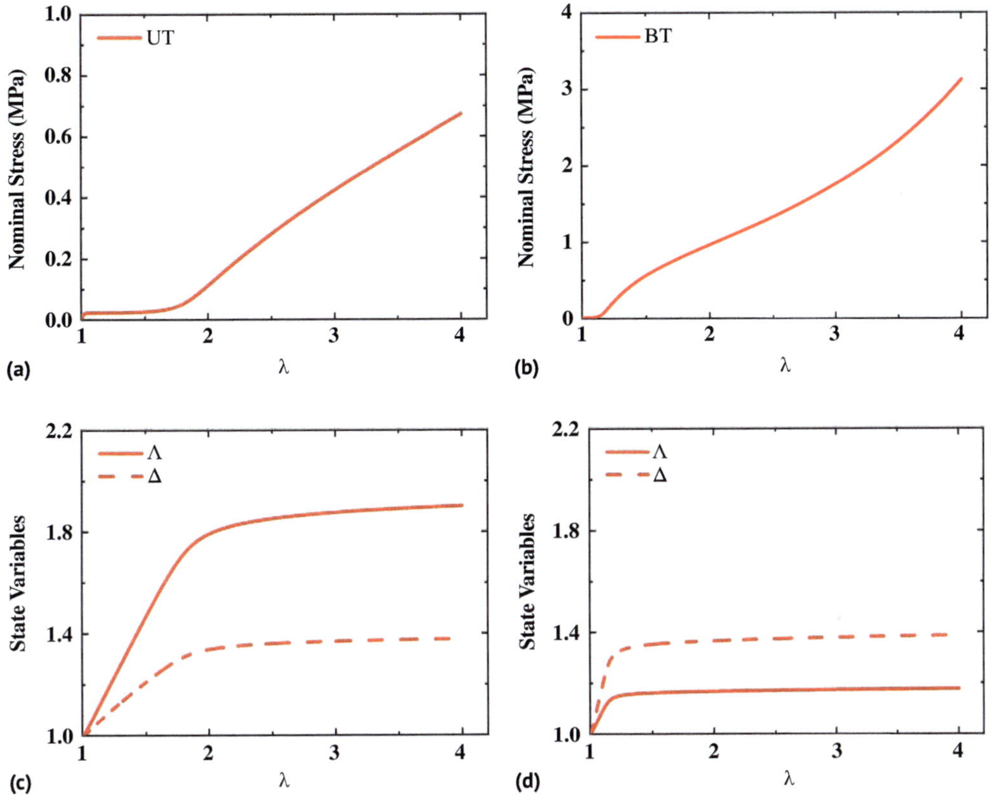

Figure 6.9 The predicted stress response in (a) uniaxial loading and (b) equibiaxial loading conditions, and the evolution of the internal variables in (c) uniaxial loading, and (d) equibiaxial loading conditions. Parameters used in the model are listed as follows: shear modulus $G = 4.94 \times 10^4$ Pa, LCE anisotropy parameter $r = 9.14$, hardening coefficient, $C = 298$ Pa, hardening exponent $k = 2$, kinetic coefficients $\nu_\Lambda = 2.18 \times 10^5$ Pa, and $\nu_\Delta = 2.18 \times 10^7$ Pa.

6.6 Conclusion

Due to the combination of entropic elasticity from the cross-linked polymer network and the orientation arising from liquid crystal rotation, LCEs exhibit unique thermal-order-mechanical coupling behaviors. This is exemplified by the thermal responsive behavior of monodomain LCEs and P-M transition behavior of polydomain LCEs. In this chapter, we present several constitutive models for both monodomain and polydomain LCEs, which effectively describe the distinctive coupling behaviors. Additionally, these models can be employed in designing self-folding structures. Nevertheless, many works still need to be performed to fully capture the complex deformation and responsive behaviors of both monodomain LCEs and polydomain LCEs, especially upon other external fields.

References

1 Gregor Skacej and Claudio Zannoni. Molecular simulations shed light on supersoft elasticity in polydomain liquid crystal elastomers. *Macromolecules*, 47(24): 8824–8832, 2014.
2 Zhi-Chao Jiang, Qing Liu, Yao-Yu Xiao, and Yue Zhao. Liquid crystal elastomers for actuation: a perspective on structure-property-function relation. *Progress in Polymer Science*, 153: 101829, 2024.
3 Daniel R Merkel, Nicholas A Traugutt, Rayshan Visvanathan, Christopher M Yakacki, and Carl P Frick. Thermomechanical properties of monodomain nematic main-chain liquid crystal elastomers. *Soft Matter*, 14(29): 6024–6036, 2018.
4 Xu Dong, Xiaoshuang Zhou, Lvzhou Li, Xiaoting Cao, Jiawei Xu, Shengping Dai, Yaoyao Jiang, Qingyue Li, Ningyi Yuan, and Jianning Ding. Monodomain liquid crystal elastomer bionic muscle fibers with excellent mechanical and actuation properties. *Iscience*, 26(4): 106357, 2023.
5 Yanjin Yao, Enjian He, Hongtu Xu, Yawen Liu, Zhijun Yang, Yen Wei, and Yan Ji. Enabling liquid crystal elastomers with tunable actuation temperature. *Nature Communications*, 14(1): 3518, 2023.
6 Daniel R Merkel, Rajib K Shaha, Christopher M Yakacki, and Carl P Frick. Mechanical energy dissipation in polydomain nematic liquid crystal elastomers in response to oscillating loading. *Polymer*, 166: 148–154, 2019.
7 Zhengxuan Wei, Peixun Wang, and Ruobing Bai. Thermomechanical coupling in polydomain liquid crystal elastomers. *Journal of Applied Mechanics*, 91(2): 021001, 2024.
8 Heino Finkelmann, Hans-J Kock, and Günther Rehage. Investigations on liquid crystalline polysiloxanes 3. Liquid crystalline elastomers—a new type of liquid crystalline material. *Die Makromolekulare Chemie, Rapid Communications*, 2(4): 317–322, 1981.
9 Katie M Herbert, Hayden E Fowler, Joselle M McCracken, Kyle R Schlafmann, Jeremy A Koch, and Timothy J White. Synthesis and alignment of liquid crystalline elastomers. *Nature Reviews Materials*, 7(1): 23–38, 2022.
10 Jonathan K Whitmer, Tyler F Roberts, Raj Shekhar, Nicholas L Abbott, and Juan J De Pablo. Modeling the polydomain-monodomain transition of liquid crystal elastomers. *Physical Review E—Statistical, Nonlinear, and Soft Matter Physics*, 87(2): 020502, 2013.
11 Devesh Mistry, Nicholas A Traugutt, Brett Sanborn, Ross H Volpe, L S Chatham, Risheng Zhou, Bo Song, Kai Yu, Kevin N Long, and Christopher M Yakacki. Soft elasticity optimises dissipation in 3D-printed liquid crystal elastomers. *Nature Communications*, 12(1): 6677, 2021.
12 Kenji Urayama, Etsuko Kohmon, Masahiro Kojima, and Toshikazu Takigawa. Polydomain-monodomain transition of randomly disordered nematic elastomers with different cross-linking histories. *Macromolecules*, 42(12): 4084–4089, 2009.
13 Pierre-Gilles De Gennes. Réflexions sur un type de polymères nématiques. *Comptes Rendus de l'Académie des Sciences. Série B*, 281: 101–103, 1975.
14 Yuzhen Chen, Alexa S Kuenstler, Ryan C Hayward, and Lihua Jin. Formation of rolls from liquid crystal elastomer bistrips. *Soft Matter*, 18(21): 4077–4089, 2022.
15 Fan Xu and Shichen Zhao. Thermal wrinkling of liquid crystal polymer shell/core spheres. *Extreme Mechanics Letters*, 40: 100860, 2020.

16 Peter Bladon, Eugene M Terentjev, and Mark Warner. Transitions and instabilities in liquid crystal elastomers. *Physical Review E*, 47(6): R3838, 1993.

17 Joachim Schätzle, Wolfgang Kaufhold, and Heino Finkelmann. Nematic elastomers: the influence of external mechanical stress on the liquid-crystalline phase behavior. *Die Makromolekulare Chemie: Macromolecular Chemistry and Physics*, 190(12): 3269–3284, 1989.

18 Helmut Finkelmann, Andreas Greve, and Mark Warner. The elastic anisotropy of nematic elastomers. *The European Physical Journal E*, 5: 281–293, 2001.

19 Lihua Jin, Zhi Zeng, and Yongzhong Huo. Thermomechanical modeling of the thermo-order–mechanical coupling behaviors in liquid crystal elastomers. *Journal of the Mechanics and Physics of Solids*, 58(11): 1907–1927, 2010.

20 Jun Wu, Yue Xiao, Yihui Zhang, and Daining Fang. A constitutive model of liquid crystal elastomers with loading-history dependence. *Journal of the Mechanics and Physics of Solids*, 174: 105258, 2023.

21 Yang Zhang, Chen Xuan, Yuefeng Jiang, and Yongzhong Huo. Continuum mechanical modeling of liquid crystal elastomers as dissipative ordered solids. *Journal of the Mechanics and Physics of Solids*, 126: 285–303, 2019.

22 Yang Zhang, Zhiyuan Zhang, and Yongzhong Huo. Nucleation and critical conditions for stripe domains in monodomain nematic elastomer sheets under uniaxial loading. *Journal of the Mechanics and Physics of Solids*, 144: 104110, 2020.

23 Yuefeng Jiang, Lihua Jin, and Yongzhong Huo. Unusual stress and strain concentration behaviors at the circular hole of a large monodomain liquid crystal elastomer sheet. *Journal of the Mechanics and Physics of Solids*, 156: 104615, 2021.

24 Isabel Kundler and Heino Finkelmann. Strain-induced director reorientation in nematic liquid single crystal elastomers. *Macromolecular Rapid Communications*, 16(9): 679–686, 1995.

25 Roberto Brighenti, Connor G McMahan, Mattia P Cosma, Arda Kotikian, Jennifer A Lewis, and Chiara Daraio. A micromechanical-based model of stimulus responsive liquid crystal elastomers. *International Journal of Solids and Structures*, 219: 92–105, 2021.

26 Sergio Conti, Antonio DeSimone, and Georg Dolzmann. Soft elastic response of stretched sheets of nematic elastomers: a numerical study. *Journal of the Mechanics and Physics of Solids*, 50(7): 1431–1451, 2002.

27 Antonio DeSimone and Luciano Teresi. Elastic energies for nematic elastomers. *The European Physical Journal E*, 29: 191–204, 2009.

28 Victoria Lee and Kaushik Bhattacharya. Actuation of cylindrical nematic elastomer balloons. *Journal of Applied Physics*, 129(11): 114701, 2021.

29 Sergei V Fridrikh and Eugene M Terentjev. Polydomain-monodomain transition in nematic elastomers. *Physical Review E*, 60(2): 1847, 1999.

30 John S Biggins, Mark Warner, and Kaushik Bhattacharya. Elasticity of polydomain liquid crystal elastomers. *Journal of the Mechanics and Physics of Solids*, 60(4): 573–590, 2012.

31 Hao Zhou and Kaushik Bhattacharya. Accelerated computational micromechanics and its application to polydomain liquid crystal elastomers. *Journal of the Mechanics and Physics of Solids*, 153: 104470, 2021.

32 Victoria Lee, Adeline Wihardja, and Kaushik Bhattacharya. A macroscopic constitutive relation for isotropic-genesis, polydomain liquid crystal elastomers. *Journal of the Mechanics and Physics of Solids*, 179: 105369, 2023.

33 Mark Warner and Eugene Michael Terentjev. *Liquid crystal elastomers*, volume 120. Oxford University Press, 2007.

34 Victoria Lee and Kaushik Bhattacharya. Universal deformations of incompressible nonlinear elasticity as applied to ideal liquid crystal elastomers. *Journal of Elasticity*, 155: 671–697, 2023.

35 L Angela Mihai, Devesh Mistry, Thomas Raistrick, Helen F Gleeson, and Alain Goriely. A mathematical model for the auxetic response of liquid crystal elastomers. *Philosophical Transactions of the Royal Society A*, 380(2234): 20210326, 2022.

36 Zheliang Wang, Ali El Hajj Chehade, Sanjay Govindjee, and Thao D Nguyen. A nonlinear viscoelasticity theory for nematic liquid crystal elastomers. *Journal of the Mechanics and Physics of Solids*, 163: 104829, 2022.

37 Gianpietro Del Piero. A rational approach to cosserat continua, with application to plate and beam theories. *Mechanics Research Communications*, 58: 97–104, 2014.

38 Weida Kang, Qian Cheng, Changyue Liu, Zhijian Wang, Dongfeng Li, and Xudong Liang. A constitutive model of monodomain liquid crystal elastomers with the thermal-mechanical-nematic order coupling. *Journal of the Mechanics and Physics of Solids*, 196: 105995, 2024.

39 Haruki Tokumoto, Hao Zhou, Asaka Takebe, Kazutaka Kamitani, Ken Kojio, Atsushi Takahara, Kaushik Bhattacharya, and Kenji Urayama. Probing the in-plane liquid-like behavior of liquid crystal elastomers. *Science Advances*, 7(25): eabe9495, 2021.

40 John S Biggins, Eugene M Terentjev, and Mark Warner. Semisoft elastic response of nematic elastomers to complex deformations. *Physical Review E—Statistical, Nonlinear, and Soft Matter Physics*, 78(4): 041704, 2008.

41 Lingrui Zhu, Mengqi He, Baihong Chen, Jin Qian, and Rui Xiao. Inflation of a polydomain nematic elastomeric membrane. *Journal of the Mechanics and Physics of Solids*, 198: 106075, 2025.

7

A Chemomechanical Model for Hydrogels

As water diffuses into a cross-linked polymer network, a hydrogel can be formed. In this chapter, we present a chemomechanical coupled model to investigate the solvent diffusion in polymer networks and the corresponding deformation behaviors of hydrogels. A thermodynamic framework is first developed for an elastomeric neutral gel. The model is then extended for two other cases. The first case is solvent diffusion and deformation of fiber-reinforced hydrogels. Due to the existence of fibers in the matrix, anisotropic responsive behaviors can be observed upon swelling. Through incorporating a free energy function of fiber deformation, the model is able to capture the self-folding behaviors of bilayer structures based on fiber-reinforced hydrogels. The second case is incorporating the glass transition mechanism into the framework. Due to the strong coupling between the glass transition and diffusion kinetics, the model successfully reproduces the main features of Case II diffusion in glassy polymers.

7.1 Introduction

Hydrogels are polymeric networks in which hydrophilic polymer chains are embedded in an aqueous medium [1]. These materials exhibit remarkable swelling behavior when exposed to aqueous environments, which endows them with the ability to undergo large deformations while maintaining structural integrity. Understanding the swelling mechanisms of hydrogels is crucial for optimizing their performance in various applications [2].

The earlier works employed two-phase models, treating the polymer network and solvent as two separate phases, to simulate the diffusion of solvents in polymer networks [3]. Though these models capture many important features of gel swelling, it is relatively difficult to achieve numerical implementation. In contrast, Hong et al. [4] formulated a chemomechanical coupled theory, which assumes a single phase for polymer networks and solvents. The model considers both large deformation of the polymer matrix and solvent diffusion. This single-phase model has been widely adopted or extended to simulate the swelling behaviors of various hydrogels [5–12]. For example, Chester and Anand [13] formulated a thermodynamically consistent model to characterize the interrelated behaviors of fluid permeation, heat transfer, and large deformation in thermally responsive elastomeric gels, which is further employed by Pantula et al. [14] to design untethered unidirectionally crawling gels. Narayan and Anand [15] developed an electro-chemo-mechanical coupled theory for polyelectrolyte gels, which can predict the transient swelling behaviors

Multi-field Modeling of Soft Active Materials: Properties and Design, First Edition. Rui Xiao.

of polyelectrolyte gels in response to different pH and ionic strength of solutions. Zimmerman et al. [16] developed a thermodynamically consistent electro-chemo-mechanical theory and computational framework to model reactive growth and remodeling in polyelectrolyte hydrogels, demonstrating its application in predicting the dynamic swelling behavior of DNA-cross-linked polymer networks.

Regardless of different types, neutral gels, thermo-responsive gels, and polyelectrolyte gels, all show isotropic swelling behaviors, which means the swelling ratio is the same in all three directions in the stress-free state. Integrating fibers into gels can induce anisotropic swelling behaviors. Gladman et al. [17] employed direct ink writing (DIW) 3D printing to fabricate composite hydrogels with controllable alignment of cellulose fibrils, which exhibit anisotropic swelling effects. The underlying mechanism is straightforward. The existence of fibers restricts the swelling along the fiber directions, while the swelling ratio is minimally influenced in other directions. In order to model these anisotropic swelling behaviors, it is necessary to incorporate the mechanism of fiber deformation, which is in fact a mature area due to modeling biological tissues. For example, Gasser et al. [18] proposed a constitutive model for arterial walls, taking into account the dispersion of collagen fiber orientation. Jaiswal et al. [19] introduced a nonsymmetric in-plane fiber distribution model for soft biological tissues. To model the responsive behaviors of fiber-reinforced hydrogels, Liu et al. [20] developed a theoretical framework with one or two families of fibers. Bosnjak et al. [21] developed a model for fiber-reinforced polymeric gels that accounts for the anisotropic diffusion of solvent within the polymer matrix. Chen et al. [22] combined experiments and numerical models to investigate the stimulus-responsive behaviors of fiber-reinforced hydrogels.

Unlike elastomers, glassy polymers exhibit distinct kinetics of penetrant sorption, often characterized by Case II or anomalous diffusion kinetics [23, 24]. Several works have been performed to investigate the diffusion in glassy polymers [25–29]. For example, Thomas and Windle [30] showed that the diffusion of the solvent is driven by osmotic pressure, while the sharp solvent front is governed by the concentration dependence of the creep response time. Rossi et al. [31] developed a phenomenological model for Case II diffusion based on the mechanism that the slow plasticization kinetics establishes an upper limit on the flux across the glassy plasticized interface. Vijalapura and Govindjee [32] proposed a robust and efficient one-dimensional numerical model to simulate Case II diffusion in glassy polymers. Wilmers and Bargmann [28] developed a fully thermodynamically consistent theory for solvent diffusion in glassy polymers, which can well capture the sharp interface observed in experiments. By combining the elasto-visco-plasticity of glassy polymers with the kinetic diffusion framework developed by Hong et al. [4], Song et al. [33] developed a model with the ability to well capture the experimentally measured diffusion behaviors in glassy polymers.

In this chapter, we aim to present a unified theoretical framework to describe solvent diffusion in different material systems, specifically, elastomers, fiber-reinforced elastomers, and glassy polymers. We first construct a thermodynamic framework to couple diffusion and deformation in polymeric gels. The explicit form of the constitutive relationship is then specified for each type of material system. A brief introduction regarding the finite element implementation is also provided. The model is then employed to simulate the swelling behavior of neutral hydrogels and fiber-reinforced hydrogels, demonstrating its ability to predict the self-folding behaviors of bilayer structures. We also show that the model is able to reproduce the main features of Case II diffusion in glassy polymers.

7.2 Thermodynamics

We first define the deformation gradient $\mathbf{F}$, which maps a point vector $\mathbf{X}$ in the undeformed configuration to a point vector $\mathbf{x}$ in the deformed configuration as

$$\mathbf{F} = \frac{\partial \mathbf{x}}{\partial \mathbf{X}}. \tag{7.1}$$

The determinant of $\mathbf{F}$ is represented as $J = \det(\mathbf{F})$.

It is typically assumed that the mechanical deformation does not change the volume of the solvent–polymer system. Rather the volume change is caused by swelling and deswelling effects. To reinforce the molecular incompressibility of the solvent, the following function is adopted:

$$1 + \upsilon C_\mathrm{s} = J, \tag{7.2}$$

where υ is the volume of single solvent molecule and C_s is the number of solvent molecules in unit volume of the reference configuration. The polymer ratio ϕ is then defined as

$$\phi = \frac{1}{1 + \upsilon C_\mathrm{s}}. \tag{7.3}$$

Mass conservation provides the following relationship between the solvent number and the flux:

$$\dot{C}_\mathrm{s} + \nabla_\mathbf{X} \cdot \mathbf{J} = 0, \tag{7.4}$$

where $\mathbf{J}$ is the flux.

The balance of the momentum can be written as

$$\nabla_\mathbf{X} \cdot \mathbf{P} + \mathbf{B} = \mathbf{0}, \tag{7.5}$$

where $\mathbf{P}$ is the first Piola–Kirchhoff stress and $\mathbf{B}$ is the body force in the reference configuration.

To simplify the diffusion process, we assume the whole diffusion and deformation processes are isothermal. Thus, heat conduction is not considered. The free energy imbalance can be represented as

$$\frac{d}{dt}\int_{\Omega_0} \Psi \mathrm{d}V \leq \int_{\Omega_0} \mathbf{B} \cdot \mathbf{V} \mathrm{d}V + \oint_{\partial\Omega_0} \mathbf{T} \cdot \mathbf{V} \mathrm{d}S - \oint_{\partial\Omega_0} \mu \mathbf{J} \cdot \mathbf{N} \mathrm{d}S, \tag{7.6}$$

where Ψ is the free energy density, $\mathbf{T}$ is the surface traction in the reference configuration, $\mathbf{V}$ is the velocity, μ is the chemical potential of the solvent, and $\mathbf{N}$ is the normal direction of the surface.

Using the integral theorem, the above equation can be simplified as

$$\frac{d}{dt}\int_{\Omega_0} \Psi \mathrm{d}V \leq \int_{\Omega_0} (\mathbf{P} : \dot{\mathbf{F}} - \mu \nabla_\mathbf{X} \cdot \mathbf{J} - \mathbf{J} \cdot \nabla_\mathbf{X} \mu) \mathrm{d}V + \int_{\Omega_0} (\nabla \mathbf{X} \cdot \mathbf{P} + \mathbf{B}) \mathrm{d}V. \tag{7.7}$$

Using the mass conservation equation (Eq. 7.4) and the balance of momentum (Eq. 7.5), this equation can be simplified as

$$\int_{\Omega_0} \dot{\Psi} \mathrm{d}V \leq \int_{\Omega_0} (\mathbf{P} : \dot{\mathbf{F}} - \mu \dot{C}_\mathrm{s} - \mathbf{J} \cdot \nabla_\mathbf{X} \mu) \mathrm{d}V. \tag{7.8}$$

The local free energy imbalance equation can then be represented as [34, 35]:

$$-\dot{\Psi} + \mathbf{P} : \dot{\mathbf{F}} + \mu \dot{C}_\mathrm{s} - \mathbf{J} \cdot \nabla_\mathbf{X} \mu \geq 0. \tag{7.9}$$

Following the method proposed in Hong et al. [36], another free energy density $\widehat{\Psi}$ is introduced through a Legendre transformation:

$$\widehat{\Psi} = \Psi - \mu C_\mathrm{s}. \tag{7.10}$$

Combining Eqs. (7.9) and (7.10) yields

$$-\dot{\widehat{\Psi}} + \mathbf{P} : \dot{\mathbf{F}} - C_\mathrm{s} \dot{\mu} - \mathbf{J} \cdot \nabla_\mathbf{X} \mu \geq 0. \tag{7.11}$$

Remark. The free energy density Ψ is a function of the number of the solvent molecules and the deformation gradient. In contrast, the free energy density $\widehat{\Psi}$ is a function of the chemical potential and the deformation gradient.

The free energy of the system has several components. The diffusion of the solvents into polymer matrix can cause a mechanical deformation of polymer networks. This further leads to an equilibrium component due to the entropic elasticity and multiple nonequilibrium components due to viscoelastic/plastic deformation processes. The mixing process of solvents can further cause a free energy change. Thus, the free energy density is represented as

$$\widehat{\Psi} = \widehat{\Psi}(\mathbf{F}, \mu, \mathbf{G}_i), \tag{7.12}$$

where $\mathbf{G}_i$ are internal variables for the viscous deformation.

The rate of $\widehat{\Psi}$ can be expanded as

$$\dot{\widehat{\Psi}} = \frac{\partial \widehat{\Psi}}{\partial \mathbf{F}} : \dot{\mathbf{F}} + \frac{\partial \widehat{\Psi}}{\partial \mu} \dot{\mu} + \sum_i \frac{\partial \widehat{\Psi}}{\partial \mathbf{G}_i} : \dot{\mathbf{G}}_i. \tag{7.13}$$

Substituting Eq. (7.13) into Eq. (7.11) gives

$$\left(\mathbf{P} - \frac{\partial \widehat{\Psi}}{\partial \mathbf{F}}\right) : \dot{\mathbf{F}} - \left(C_\mathrm{s} + \frac{\partial \widehat{\Psi}}{\partial \mu}\right) \dot{\mu} - \sum_i \frac{\partial \widehat{\Psi}}{\partial \mathbf{G}_i} : \dot{\mathbf{G}}_i - \mathbf{J} \cdot \nabla_\mathbf{X} \mu \geq 0. \tag{7.14}$$

At the equilibrium state, there is no free energy dissipation, which requires the following relationship to be held:

$$\mathbf{P} = \frac{\partial \widehat{\Psi}}{\partial \mathbf{F}}, \tag{7.15a}$$

$$C_\mathrm{s} = -\frac{\partial \widehat{\Psi}}{\partial \mu}. \tag{7.15b}$$

The Cauchy stress is related to the first Piola–Kirchhoff stress as

$$\boldsymbol{\sigma} = \frac{1}{J}\mathbf{P}\mathbf{F}^{\mathrm{T}}. \tag{7.16}$$

Combining Eqs. (7.14) and (7.15), the second law of thermodynamics is reduced to

$$-\sum_i \frac{\partial \widehat{\Psi}}{\partial \mathbf{G}_i} : \dot{\mathbf{G}}_i - \mathbf{J} \cdot \nabla_{\mathbf{X}}\mu \geq 0. \tag{7.17}$$

Since the solvent diffusion process and viscoelastic/plastic deformation can be separate processes, a stronger form of free energy imbalance can be rewritten as follows to satisfy Eq. (7.17):

$$-\sum_i \frac{\partial \widehat{\Psi}}{\partial \mathbf{G}_i} : \dot{\mathbf{G}}_i \geq 0, \tag{7.18a}$$

$$-\ \mathbf{J} \cdot \nabla_{\mathbf{X}}\mu \geq 0. \tag{7.18b}$$

In this part, we first specify the explicit equation of the solvent diffusion to satisfy Eq. (7.18b), while we leave the requirement of the thermodynamics for viscous dissipation in the next section.

For the diffusion process, the following equation is adopted to satisfy Eq. (7.18b) [4, 37]:

$$\mathbf{J} = -\mathbf{M}\nabla\mu, \tag{7.19}$$

where $\mathbf{M}$ is the positive-definite mobility tensor.

Specifically, the framework proposed in Hong et al. [4] is adopted for the kinetics of solvent diffusion. In the deformed configuration, the solvent flux is related to the gradient of the chemical potential as

$$\mathbf{j} = -\frac{Dc_{\mathrm{s}}}{kT}\nabla_{\mathbf{x}}\mu, \tag{7.20}$$

where j and c_{s} are the flux and solvent molecular number in the deformed configuration, k is the Boltzmann constant, T is the temperature, and D is the diffusion coefficient. It should be noted that the diffusion coefficient may depend on the solvent concentration.

The flux in the deformed and reference configuration has the following relationship: $\mathbf{j}\mathrm{d}\mathbf{s} = \mathbf{J}\mathrm{d}\mathbf{S}$. Through using Nanson's formula $\mathrm{d}\mathbf{s} = J\mathbf{F}^{-\mathrm{T}}\mathrm{d}\mathbf{S}$, we can obtain the following relation:

$$\mathbf{j} = \frac{\mathbf{F}}{J}\mathbf{J}. \tag{7.21}$$

The nominal solvent number C_{s} is related to the solvent number in the deformed configuration as

$$c_{\mathrm{s}} = \frac{C_{\mathrm{s}}}{J}. \tag{7.22}$$

The gradient of chemical potential can be presented in the reference configuration as follows:

$$\nabla_{\mathbf{x}}\mu = \mathbf{F}^{-\mathrm{T}}\nabla_{\mathbf{X}}\mu. \tag{7.23}$$

Combining Eqs. (7.20)–(7.23), the following equation is obtained for the diffusion in the reference state:

$$\mathbf{J} = -\frac{DC_s}{kT}\mathbf{F}^{-1}\nabla_X \mu \mathbf{F}^{-1}. \tag{7.24}$$

To obtain the constitutive relationship, the explicit form of the free energy density needs to be specified, which will be discussed in the following section.

7.3 Constitutive Relations

In this section, we will specify the chemomechanical coupled model to describe the solvent diffusion in three different polymeric systems, namely, elastomers, fiber-reinforced elastomers, and glassy polymers.

7.3.1 Neutral Gels

For an elastomeric system, the diffusion of solvents causes a change in volume. The free energy change has two contributions, the stretching of polymer chains (Ψ_{net}) and the mixing of solvent and polymer networks (Ψ_{mix}), which can be represented as

$$\Psi = \Psi_{net} + \Psi_{mix}. \tag{7.25}$$

The Gaussian type of free energy density is adopted for stretching polymer chains:

$$\Psi_{net} = \frac{G_{net}}{2}\left[\mathrm{I}_1 - 3 - 2\ln(J)\right], \tag{7.26}$$

where G_{net} is the shear modulus of the polymer matrix in the dry state, I_1 is the first invariant of the right Cauchy–Green strain tensor $\mathbf{C} = \mathbf{F}^{\mathrm{T}}\mathbf{F}$. It is also typically written as $G_{net} = nKT$, where n is the number density of polymer chains in the reference configuration.

The free energy density of the polymer–solvent mixture can be expressed as [38, 39]

$$\Psi_{mix} = -\frac{kT}{\upsilon}\left[\upsilon C_s \log\left(1 + \frac{1}{\upsilon C_s}\right) + \frac{\chi}{1 + \upsilon C_s}\right], \tag{7.27}$$

where χ is the dimensionless parameter that describes the enthalpy change when the network and solvents mix in gels. In this equation, the first term on the right represents the entropy change caused by mixing and the second term represents the enthalpy change due to mixing.

Using the relation in Eq. (7.2), the equation can be rewritten as

$$\Psi_{mix} = -\frac{kT}{\upsilon}\left[(J-1)\log(\frac{J}{J-1}) + \frac{\chi}{J}\right]. \tag{7.28}$$

Combining Eqs. (7.10), (7.25), (7.26), and (7.28), the free energy density of the hydrogel can be obtained:

$$\begin{aligned}\tilde{\Psi}(\mathbf{F}, \mu) = \frac{G_{net}}{2}\left[\mathrm{I}_1 - 3 - 2\log(J)\right] - \frac{kT}{\upsilon}\left[(J-1)\log\left(\frac{J}{J-1}\right) + \frac{\chi}{J}\right] \\ -\frac{\mu}{\upsilon}(J-1).\end{aligned} \tag{7.29}$$

Substituting Eq. (7.29) into Eqs. (7.15a) and (7.16) yields the explicit expression of the stress as

$$\mathbf{P} = G_{\text{net}}\left(\mathbf{F} - \mathbf{F}^{-T}\right) + \frac{kT}{v}\left[J\log\left(1 - \frac{1}{J}\right) + 1 + \frac{\chi}{J}\right]\mathbf{F}^{-T} - \frac{\mu}{v}J\mathbf{F}^{-T}, \tag{7.30a}$$

$$\sigma = \frac{1}{J}G_{\text{net}}(\mathbf{b} - \mathbf{I}) + \frac{1}{J}\frac{kT}{v}\left[J\log\left(1 - \frac{1}{J}\right) + 1 + \frac{\chi}{J}\right]\mathbf{I} - \frac{\mu}{v}\mathbf{I}. \tag{7.30b}$$

Remark. As shown in Hong et al. [4], the chemical potential can be represented as $\mu = \frac{kT}{v}\left[\ln\left(1 - \frac{1}{J}\right) + \frac{1}{J} + \frac{\chi}{J^2}\right] - pv$, where p is the hydrostatic pressure. Thus, the two last terms on the right-hand side of Eq. (7.30b) reduces to $p\mathbf{I}$.

7.3.2 Fiber-reinforced Gels

As fiber is introduced into the polymer matrix, the swelling is restricted in the fiber direction. The central task is including the contribution of the fiber deformation into the free energy. However, fiber-reinforced gels are mostly fabricated using DIW 3D printing. The stress-free state of fibers is not the same as dry polymers. Rather, it is naturally related to an initial swelling state. Thus, we adopt this initial swelling state as the reference state with the initial swelling ratio λ_0 in each direction.

The total free energy density can then be represented as

$$\Psi^{\text{f}} = \Psi^{\text{f}}_{\text{net}} + \Psi^{\text{f}}_{\text{fiber}} + \Psi^{\text{f}}_{\text{mix}}. \tag{7.31}$$

To distinguish from the free energy density of the neutral gels, a superscript "f" is added.

The free energy density of the matrix is represented as

$$\Psi^{\text{f}}_{\text{net}} = \frac{1}{\lambda_0^3}\frac{G_{\text{net}}}{2}\left[\lambda_0^2 \text{I}_1 - 3 - 2\ln(\lambda_0^3 J)\right]. \tag{7.32}$$

Many classic models [18, 40, 41] can be adopted for the free energy caused by fiber deformation. Here, we use the following form:

$$\Psi^{\text{f}}_{\text{fiber}} = \frac{G_{\text{fiber}}}{2k_2}\left[\exp\left(k_2(I_4 - 1)^2\right) - 1\right], \tag{7.33}$$

where G_{fiber} is the shear modulus of the fiber, and k_2 controls the strain stiffening effect. The fiber distribution is assumed to be perfectly aligned in the direction $\mathbf{a}_0$, with $\mathbf{a}_0$ being a unit vector. The square of the stretch ratio of this unit vector upon the deformation gradient can be written as

$$I_4 = \mathbf{a}_0 \cdot \mathbf{C}\mathbf{a}_0. \tag{7.34}$$

The mixing free energy density can be represented in the following form:

$$\Psi^{\text{f}}_{\text{mix}} = -\frac{1}{\lambda_0^3}\frac{kT}{v}\left[(\lambda_0^3 J - 1)\log\left(\frac{\lambda_0^3 J}{\lambda_0^3 J - 1}\right) + \frac{\chi}{\lambda_0^3 J}\right]. \tag{7.35}$$

Combining Eqs. (7.10), (7.31), (7.32), and (7.35) yields the following form of free energy density:

$$\begin{aligned}\widetilde{\Psi}^{\mathrm{f}}(\mathbf{F},\mu) &= \frac{1}{\lambda_0^3}\frac{G_{\mathrm{net}}}{2}\left[\lambda_0^2 \mathrm{I}_1 - 3 - 2\ln(\lambda_0^3 J)\right] + \frac{G_{\mathrm{fiber}}}{2k_2}\left[\exp\left(k_2(I_4-1)^2\right)-1\right] \\ &\quad - \frac{1}{\lambda_0^3}\frac{kT}{v}\left[(\lambda_0^3 J - 1)\log\left(\frac{\lambda_0^3 J}{\lambda_0^3 J - 1}\right) + \frac{\chi}{\lambda_0^3 J}\right] - \frac{1}{\lambda_0^3}\frac{\mu}{v}\left(\lambda_0^3 J - 1\right).\end{aligned} \tag{7.36}$$

Substituting Eq. (7.36) into Eqs. (7.15a) and (7.16) leads to the following explicit form for the first Piola–Kirchhoff stress and the Cauchy stress:

$$\begin{aligned}\mathbf{P} &= \frac{1}{\lambda_0^3}G_{\mathrm{net}}\left(\lambda_0^2\mathbf{F} - \mathbf{F}^{-\mathrm{T}}\right) + 2G_{\mathrm{fiber}}\exp\left(k_2(I_4-1)^2\right)\mathbf{F}\mathbf{a_0}\otimes\mathbf{a_0} \\ &\quad + \frac{1}{\lambda_0^3}\frac{kT}{v}\left[\lambda_0^3 J\log\left(1-\frac{1}{\lambda_0^3 J}\right) + 1 + \frac{\chi}{\lambda_0^3 J}\right]\mathbf{F}^{-\mathrm{T}} - \frac{\mu}{v}J\mathbf{F}^{-\mathrm{T}},\end{aligned} \tag{7.37a}$$

$$\begin{aligned}\boldsymbol{\sigma} &= \frac{1}{J}G_{\mathrm{net}}\left(\frac{\mathbf{b}}{\lambda_0} - \frac{\mathbf{I}}{\lambda_0^3}\right) + \frac{2}{J}G_{\mathrm{fiber}}\exp\left(k_2(I_4-1)^2\right)\mathbf{a_n}\otimes\mathbf{a_n} \\ &\quad + \frac{1}{J}\frac{kT}{v}\left[J\log\left(1-\frac{1}{\lambda_0^3 J}\right) + \frac{1}{\lambda_0^3} + \frac{\chi}{\lambda_0^6 J}\right]\mathbf{I} - \frac{\mu}{v}\mathbf{I},\end{aligned} \tag{7.37b}$$

where $\mathbf{a_n} = \mathbf{F}\mathbf{a_0}$ is the new position of the vector $\mathbf{a_0}$ in the deformed configuration.

7.3.3 Diffusion in Glassy Elastomers

The viscous deformation process is an energy dissipation process. To model such processes, we further decompose the total free energy density into a time-independent equilibrium component $\Psi^{\mathrm{g}}_{\mathrm{net}}$, multiple time-dependent components Ψ^{g}_i, and a mixing part $\Psi^{\mathrm{g}}_{\mathrm{mix}}$

$$\Psi^{\mathrm{g}} = \Psi^{\mathrm{g}}_{\mathrm{net}} + \sum_i \Psi^{\mathrm{g}}_i + \Psi^{\mathrm{g}}_{\mathrm{mix}}. \tag{7.38}$$

Again, a superscript "g" is added to distinct it from the other two cases.

The explicit form of $\Psi^{\mathrm{g}}_{\mathrm{net}}$ is set the same as Eq. (7.26), while the explicit form of $\Psi^{\mathrm{g}}_{\mathrm{mix}}$ is set the same as Eq. (7.27).

Based on Eqs. (7.10), (7.15a), (7.16), and (7.38), the corresponding nonequilibrium stress terms are defined as

$$\mathbf{P}_i^{\mathrm{neq}} = \frac{\partial \Psi^{\mathrm{g}}_i}{\partial \mathbf{F}}, \tag{7.39a}$$

$$\boldsymbol{\sigma}_i^{\mathrm{neq}} = \frac{1}{J}\frac{\partial \Psi^{\mathrm{g}}_i}{\partial \mathbf{F}}\mathbf{F}^{\mathrm{T}}. \tag{7.39b}$$

Motivated by the small strain rheological model, we propose the following evolution equation for the nonequilibrium stress as [42, 43]:

$$\boldsymbol{\sigma}_i^{\mathrm{neq}}(t) = \int_{-\infty}^{t} \beta_i \exp\left(-\frac{t-s}{\tau_i(s)}\right)\frac{\mathrm{d}}{\mathrm{d}s}\left(\frac{1}{J}\frac{\partial \Psi^{\mathrm{g}}_{\mathrm{net}}}{\partial \mathbf{F}}\mathbf{F}^{\mathrm{T}}(s)\right)\mathrm{d}s, \tag{7.40}$$

where β_i represents the ratio between the ith nonequilibrium modulus and the equilibrium modulus, and τ_i is the relaxation time. The relaxation time depends on temperature and the solvent/polymer ratio.

We further define $\boldsymbol{\sigma}_{i0}^{\text{neq}}(t) = \beta_i \frac{\partial \Psi_{\text{net}}^{\text{g}}}{\partial \mathbf{F}} \mathbf{F}^{\text{T}}(t)$, with the following form:

$$\boldsymbol{\sigma}_{i0}^{\text{neq}} = \frac{1}{J} \beta_i G_{\text{net}} (\mathbf{b} - \mathbf{I}) . \tag{7.41}$$

The integral form of the stress evolution equals the following local differential equation:

$$\dot{\boldsymbol{\sigma}}_i^{\text{neq}}(t) + \frac{\boldsymbol{\sigma}_i^{\text{neq}}(t)}{\tau_i(t)} = \dot{\boldsymbol{\sigma}}_{i0}^{\text{neq}}(t) . \tag{7.42}$$

Remark. The viscoelastic/plastic processes are required to satisfy the entropy inequality. Adopting Eq. (7.42) for relaxation processes cannot be directly proved to satisfy the second law of thermodynamics. However, as discussed in Christensen [42], the entropy inequality can be strictly proved in the small strain condition.

Compared with the neutral gels, the stress terms will have extra contributions from the nonequilibrium processes, represented as

$$\mathbf{P} = G_{\text{net}} \left(\mathbf{F} - \mathbf{F}^{\text{-T}} \right) + \sum_i \mathbf{P}_i^{\text{neq}} + \frac{kT}{\upsilon} \left[J \log \left(1 - \frac{1}{J} \right) + 1 + \frac{\chi}{J} \right] \mathbf{F}^{\text{-T}} - \frac{\mu}{\upsilon} J \mathbf{F}^{\text{-T}}, \tag{7.43a}$$

$$\boldsymbol{\sigma} = \frac{1}{J} G_{\text{net}} (\mathbf{b} - \mathbf{I}) + \sum_i \boldsymbol{\sigma}_i^{\text{neq}} + \frac{1}{J} \frac{kT}{\upsilon} \left[J \log \left(1 - \frac{1}{J} \right) + 1 + \frac{\chi}{J} \right] \mathbf{I} - \frac{\mu}{\upsilon} \mathbf{I}. \tag{7.43b}$$

For amorphous polymers, especially when the temperature is around or below the glass transition temperature, the relaxation time and diffusion coefficient both strongly depend on the solvent concentration. The value can be changed by several orders of magnitude with the absorption of a small amount of solvents [44, 45]. Thus, it is essential to define the explicit form for the relaxation time and diffusion coefficient.

To model the glass transition, we adopt the classic free volume theory. Here, the free volume is extended to include solvent concentration dependence

$$f = f_p + (1 - \phi) f_s^{\infty} \tag{7.44}$$

where f is the total free volume, f_p is the free volume of the pure polymer, and f_s^{∞} is the maximum free volume change caused by the solvent.

The Doolittle equation is adopted to express the relation between the relaxation time and free volume:

$$\tau_i = A_i \exp \left(\frac{B}{f} \right), \tag{7.45}$$

where A_i and B are model parameters.

This equation is rewritten by considering the reference relaxation time $\tau_{\text{ref}} = A_i \exp (B/f_p)$,

$$\tau_i = \tau_i^{\text{ref}} \exp \left(\frac{B}{f} - \frac{B}{f_p} \right) . \tag{7.46}$$

Several forms have been proposed to model the dependence of the diffusion coefficient on the solvent concentration. Here, we adopt the form presented in Vijalapura and Govindjee [32]:

$$D = D_0\left[1 + \alpha\psi^2\right], \tag{7.47}$$

where D_0 is the reference diffusion coefficient and α is a parameter representing the increase in the diffusion speed due to the increase in the solvent fraction. The term ψ is the normalized solvent concentration, defined as $\psi = \frac{(J-1)/J}{(J^{eq}-1)/J^{eq}}$, where J^{eq} is the determinant in the final equilibrium state. Thus, the dry state gives $\psi = 0$, while the final equilibrium swelling state yields $\psi = 1$. This indicates that the diffusion coefficient increases from D_0 in the dry state to $(1+\alpha)D_0$ in the fully swollen state.

7.3.4 Numerical Method

The above model is implemented into COMSOL Multiphysics for finite element analysis. Here, we briefly describe the procedures. The balance of momentum in Eq. (7.5) is the strong form for the mechanical process. The corresponding weak form can be represented as

$$\int \nabla_{\mathbf{X}} \cdot \mathbf{P} \cdot \mathbf{w}_1 \mathrm{dV} + \int \mathbf{B} \cdot \mathbf{w}_1 \mathrm{dV} = 0, \tag{7.48}$$

where $\mathbf{w}_1$ is the weighting field.

This equation can be rewritten as

$$\int \mathbf{P} : \nabla_{\mathbf{X}} \mathbf{w}_1 \mathrm{dV} = \int \mathbf{B} \cdot \mathbf{w}_1 \mathrm{dV} + \int_S \hat{\mathbf{T}} \cdot \mathbf{w}_1 \mathrm{dS}, \tag{7.49}$$

where $\hat{\mathbf{T}}$ is the traction force in the reference configuration.

For the diffusion process, the mass conservation is the strong form. We first rewrite Eq. (7.4) by replacing C_s using J as

$$\frac{\dot{J}}{v} + \nabla_{\mathbf{X}} \cdot \mathbf{J} = 0. \tag{7.50}$$

The weak form can then be written as

$$\int \frac{\dot{J}}{v} w_2 \mathrm{dV} + \int \nabla_{\mathbf{X}} \cdot \mathbf{J} w_2 \mathrm{dV} = 0, \tag{7.51}$$

where w_2 is the weighting function.

We further rewrite this equation as

$$\int \frac{\dot{J}}{v} w_2 \mathrm{dV} + \int \mathbf{J} \cdot \nabla_{\mathbf{X}} w_2 \mathrm{dV} = \int_S \hat{J} w_2 \mathrm{dS}, \tag{7.52}$$

where $\hat{J}$ is the fluid flux on the surface in the reference configuration.

The fluid flux in the reference configuration $\mathbf{J}$ can be evaluated from a modified form of Eq. (7.20) as

$$\mathbf{J} = -\frac{D(J-1)}{vkT} \mathbf{F}^{-1} \nabla_{\mathbf{X}} \mu \mathbf{F}^{-1}. \tag{7.53}$$

Thus, Eq. (7.52) can be rewritten as

$$\int \frac{\dot{J}}{\upsilon} w_2 \mathrm{d}V + \int \frac{D(J-1)}{\upsilon kT} \mathbf{F}^{-1}\nabla_{\mathbf{X}}\mu \cdot \mathbf{F}^{-1}\nabla_{\mathbf{X}} w_2 \mathrm{d}V = \int_S \hat{J} w_2 \mathrm{d}S. \tag{7.54}$$

The nonequilibrium stress is adopted as internal variables. Using the method proposed in Zhong et al. [43], we define the following internal variables as

$$\mathbf{q}_i = \boldsymbol{\sigma}_{i0}^{\mathrm{neq}} - \boldsymbol{\sigma}_i^{\mathrm{neq}}. \tag{7.55}$$

Equation (7.42) is then represented as

$$\dot{\mathbf{q}}_i(t) + \frac{\mathbf{q}_i(t)}{\tau_i(t)} = \boldsymbol{\sigma}_{i0}^{\mathrm{neq}}(t). \tag{7.56}$$

For each time step, $\mathbf{q}_i$ is evaluated using Eq. (7.56), which further gives the value for $\boldsymbol{\sigma}_i^{\mathrm{neq}}$. The nonequilibrium Cauchy stress needs to be pulled back to the reference configuration to obtain the nonequilibrium first Piola–Kirchhoff stress using Eq. (7.39b). The nonequilibrium first Piola–Kirchhoff stress will be updated to the Piola–Kirchhoff stress. The weak form of Eq. (7.49) is solved in the Solid Mechanics Module of COMSOL, while Eq. (7.54) is solved in the weak form.

7.4 Results

In this part, we investigate the performance of the above models. A normalized model is adopted. Specifically, the modulus is normalized by kT/υ as $\overline{G}_{\mathrm{net}} = G_{\mathrm{net}}\upsilon/(kT)$ and $\overline{G}_{\mathrm{fiber}} = G_{\mathrm{fiber}}\upsilon/(kT)$. The chemical potential is normalized by $\bar{\mu} = \mu/kT$.

7.4.1 Neutral Gels and Fiber-reinforced Gels

It can be seen that the model for neutral gels is a specific case of fiber-reinforced gels when the shear modulus of fibers is set as 0. Thus, we present the results for the neutral gels and fiber-reinforced gels together in this part. The model parameters are set as: $\overline{G}_{\mathrm{net}} = 0.001$, $\chi = 0.6$, $\lambda_0 = 1.5$, and $k_2 = 1$. The diffusion constant is set as 1.

From Eq. (7.37b), we can obtain the swelling ratio as a function of the normalized chemical potential. Specifically, the principal stress $\sigma_i = 0$ for $i = 1, 2, 3$ gives three equations. Once the chemical potential is specified, the three principal stretches can be calculated. As shown in Figure 7.1, the swelling ratio is the same for the three principal directions for neutral gels without fibers. The swelling ratio increases rapidly when the chemical potential approaches zero. We then investigate the swelling behaviors of fiber-reinforced hydrogels with the modulus of the fibers set the same as the matrix as $\overline{G}_{\mathrm{fiber}} = 0.001$. The fiber is assumed to align in $\mathbf{e}_1$ direction. The swelling ratio is much smaller along the fiber direction than the other two directions, while the swelling ratio is slightly larger in the perpendicular directions than neutral gels without fibers. A further increase in the modulus of fibers can reduce the swelling ratio along the fiber directions [46].

Since the model incorporates the diffusion process, it can be employed to simulate the shape change with time upon swelling. A rectangular specimen with a dimension of $10 \times 10 \times 1$ is adopted. Due to the symmetric condition, only one eighth of the whole specimen is simulated. As shown in Figure 7.2, the top surface and two side faces are assumed to

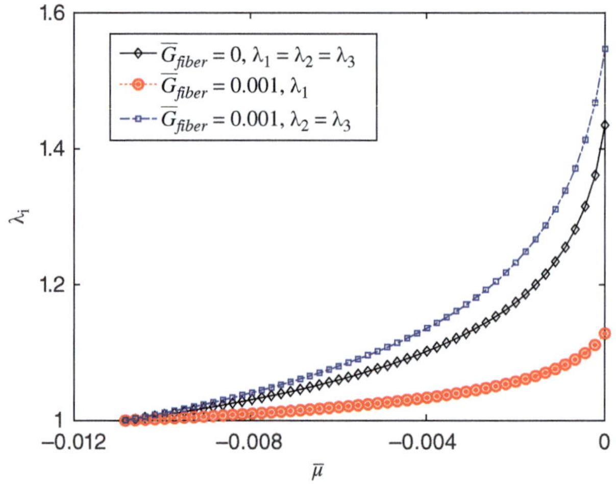

Figure 7.1 Swelling ratio in the length as a function of normalized chemical potential in the free swelling state for neutral gels and fiber-reinforced gels.

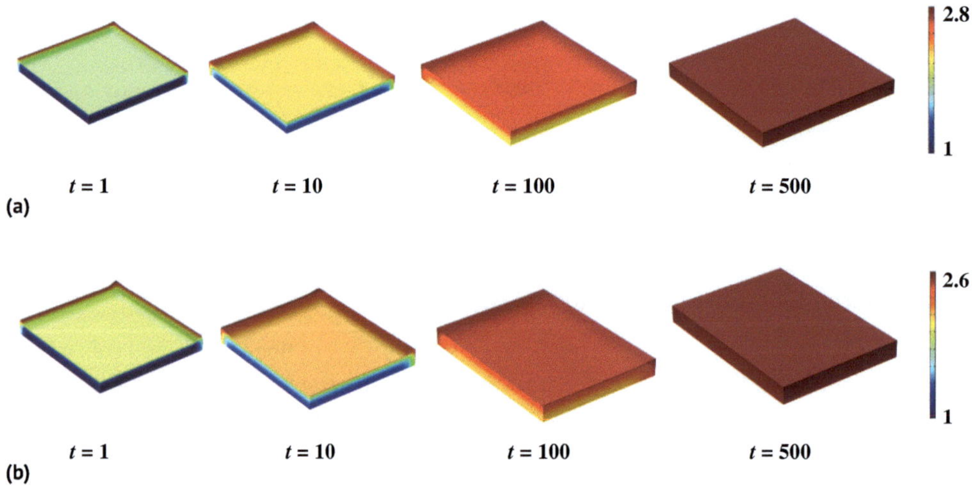

Figure 7.2 The kinetic diffusion behavior of a thin hydrogel film (a) without fibers and (b) with fibers. The contour corresponds to the determinant of the deformation gradient *J*. *Source:* Dai et al./ with permission of World Scientific Publishing.

be in contact with water. The chemical potential is ramped from the initial value to 0 within a time of 0.01. The initial value of the chemical potential can be calculated using Eq. (7.37b) through the stress-free condition together with $\lambda_0 = 1.5$. As shown in Figure 7.2, the outer part swells first because more areas make contact with water. With time, a more homogeneous swelling state can be reached. The anisotropic swelling behaviors of fiber-reinforced hydrogels can be clearly observed compared with the isotropic swelling state of neutral gels without fibers. The final swelling ratio is slightly smaller for fiber-reinforced gels compared with neutral gels. This is because a relatively low modulus of fibers is adopted here.

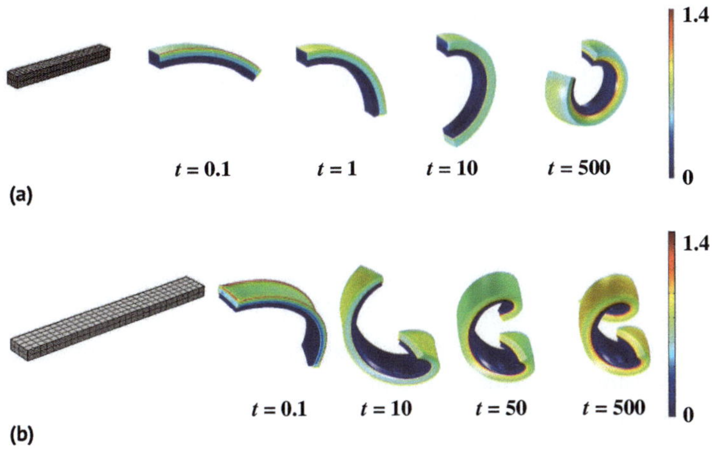

Figure 7.3 The shape morphing behaviors of bilayer stripes with active layers using hydrogels (a) without fibers and (b) with fibers. The contour corresponds to the determinant of the deformation gradient J. *Source:* Dai et al./with permission of World Scientific Publishing.

Figure 7.3 shows the swelling and bending of a bilayer structure. The upper layer of the bilayer structure is a hydrogel, while the lower layer is a passive polymer that does not swell in water. The passive layer is modeled using a neo-Hookean model with a normalized shear modulus of 0.001. The size of each layer is set as $20 \times 2.5 \times 0.5$. For gels without fibers, only one-fourth of the structure is simulated. As shown in Figure 7.3, with an increase in the swelling time, the volume of the hydrogel gradually expands, causing the whole structure to bend progressively. A similar design has also been demonstrated in Chester et al. [47]. Different from the framework in Chester et al. [47], we have used the Legendre transformation to change the state variable from the number of solvents to the chemical potential. This leads to a feasible numerical implementation procedure.

Fiber-reinforced hydrogels are then adopted as the active layer. The direction of the fiber is assigned as 45° along the length direction. As shown, the structure transits to a helix shape with swelling. Thus, introducing fibers can induce another degree of freedom in design through controlling the distribution of fibers. We have also carried out the corresponding experiments to validate the above model in Xu et al. [48]. The results show that the model can well reproduce the transient shape change of fiber-reinforced hydrogels and corresponding structures. It is also revealed that the deformation modes change from a helicoid to a spiral ribbon with the increasing width-to-thickness ratios. Interested readers can refer to Xu et al. [48] for more details.

7.4.2 Diffusion in Glassy Polymers

Case II diffusion can occur when solvents diffuse into glassy polymers. It is characterized by a sharp moving front, a distinct boundary that separates the swollen region and the dry region. The moving velocity is almost constant. In the swollen region, this is only a small variation of the swelling ratio. The mass increases linearly with time. These features are totally different from the solvent diffusion in elastomers. To fully capture these behaviors, two additional important factors need to be considered. First, the solvent-induced glass transition needs to be fully characterized. We use the free volume theory to model this effect. Second, the diffusion coefficient also increases by several orders from the dry state to a fully swollen state. We have also adopted a phenomenological equation (Eq. 7.47) to capture this effect.

We still use the normalized model to qualitatively reproduce Case II diffusion. Thus, all the modulus is scaled by kT/v and the chemical potential is scaled by kT. To be simple, only one relaxation process is employed. The parameters are set as follows: $\overline{G}_{\text{net}}^{g} = 0.01$, $\beta = 150$, $\chi = 1.368$, $\tau = 800$, $f_p = 0.025$, $f_s^{\infty} = 0.025$, $B = 1.0$, $D_0 = 0.001$, and $\alpha = 1000$. A rectangular specimen with a dimension of $10 \times 10 \times 1$ is employed. It is further assumed that only the top surface makes contact with water.

The solvent diffusion in these material systems is shown in Figure 7.4. Several distinct characteristics of Case II diffusion can be observed. Firstly, a sharp diffusion front develops with time. This is in contrast to Fickian diffusion, where the concentration profile is typically more gradual. Secondly, a small increase in the volume ratio J is observed ahead of the sharp front. This phenomenon has been previously described as a "Fickian precursor." Finally, the concentration gradient behind the sharp front is found to be negligible. This can also be observed from the finite element results as shown in Figure 7.5a. The swelling degree Q, defined as the average of $J-1$ over the whole specimen, as a function of

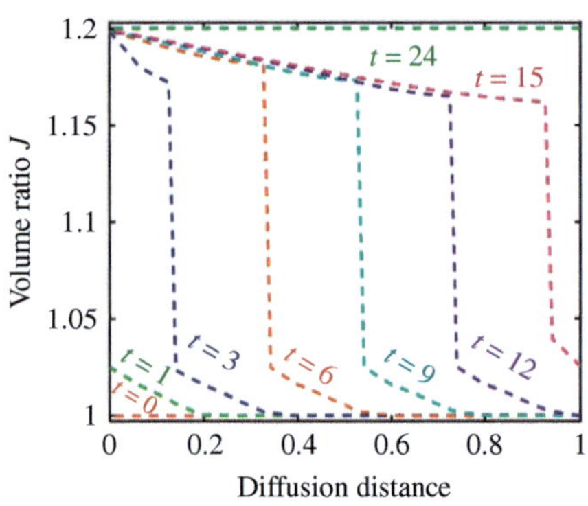

Figure 7.4 The distribution of the volume ratio J with respect to the diffusion distance at different time.

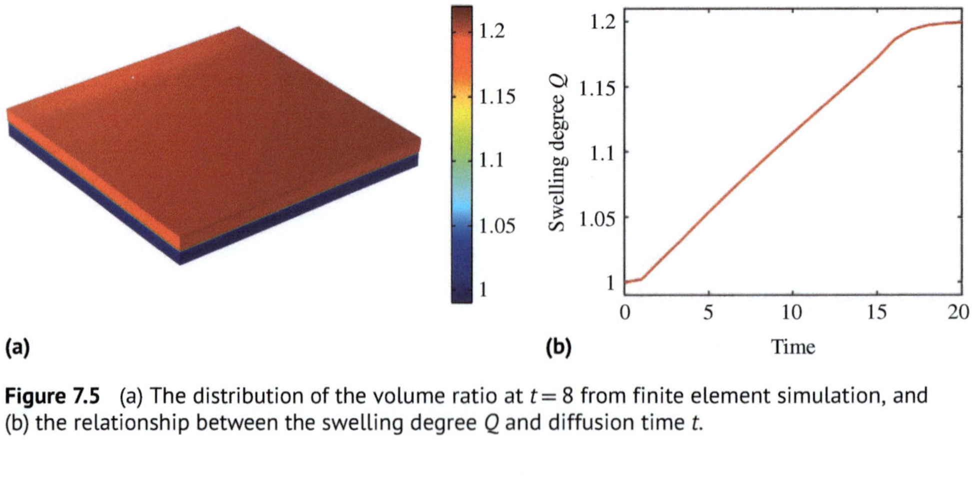

Figure 7.5 (a) The distribution of the volume ratio at $t = 8$ from finite element simulation, and (b) the relationship between the swelling degree Q and diffusion time t.

time is shown in Figure 7.5b. It is observed that the swelling degree changes linearly with diffusion time showing a clear induction time. These are all the main features shown in experiments [49].

7.5 Conclusion

Hydrogels show a considerable difference in responsive behaviors compared with other material systems introduced in previous chapters. The volume of hydrogels can be significantly changed due to swelling, while the volume only shows limited change for other material systems. Developing theoretical models needs to consider this important feature. In this chapter, we develop a thermodynamic framework to couple the kinetics of solvent diffusion and finite deformation behaviors of hydrogels. The model is then specified for three cases. The first case is the simplest gel systems neutral gels, while the second case is fiber-reinforced gels exhibiting an anisotropic swelling behavior. We also apply the model to investigate the self-folding behaviors of bilayer structures with a hydrogel layer and a passive layer. It is shown that fiber-reinforced gels provide another design freedom for stimulus-responsive systems through controlling the fiber directions. The theoretical framework is then extended to simulate Case II diffusion in glassy polymers through incorporating the solvent-induced glassy-rubbery transition. It is shown that the model is able to capture the important features of Case II diffusion. We hope that this chapter can promote the future development of modeling various types of gels and simulating smart structures based on hydrogels.

References

1 Junjie Liu, Shaoxing Qu, Zhigang Suo, and Wei Yang. Functional hydrogel coatings. *National Science Review*, 8(2): nwaa254, 2021.

2 Rong Huang, Shoujing Zheng, Zishun Liu, and Teng Yong Ng. Recent advances of the constitutive models of smart materials — hydrogels and shape memory polymers. *International Journal of Applied Mechanics*, 12(02): 2050014, 2020.

3 Yong Li and Toyoichi Tanaka. Kinetics of swelling and shrinking of gels. *The Journal of Chemical Physics*, 92(2): 1365–1371, 1990.

4 Wei Hong, Xuanhe Zhao, Jinxiong Zhou, and Zhigang Suo. A theory of coupled diffusion and large deformation in polymeric gels. *Journal of the Mechanics and Physics of Solids*, 56(5): 1779–1793, 2008.

5 Yiheng Xue, Zishun Liu, and Junuthulla N Reddy. A three-fields coupled numerical framework for transient deformation of thermo-sensitive hydrogel. *International Journal for Numerical Methods in Engineering*, 125(18): e7550, 2024.

6 Jianying Hu, William Toh, Teng Yong Ng, Nan Jiang, Liangsong Zeng, Jianke Du, and Zishun Liu. Inhomogeneous large deformation study on magneto-thermal sensitive hydrogels. *International Journal of Applied Mechanics*, 13(05): 2150053, 2021.

7 Romain Marcombe, Shengqiang Cai, Wei Hong, Xuanhe Zhao, Yuri Lapusta, and Zhigang Suo. A theory of constrained swelling of a pH-sensitive hydrogel. *Soft Matter*, 6(4): 784–793, 2010.

8 Yalin Yu, Chad M Landis, and Rui Huang. Salt-induced swelling and volume phase transition of polyelectrolyte gels. *Journal of Applied Mechanics*, 84(5): 051005, 2017.

9 Zishun Liu, William Toh, and Teng Yong Ng. Advances in mechanics of soft materials: a review of large deformation behavior of hydrogels. *International Journal of Applied Mechanics*, 7(05): 1530001, 2015a.

10 Aleksey D Drozdov, Jesper deClaville Christiansen, and Catalina-Gabriela Sanporean. Inhomogeneous swelling of pH-responsive gels. *International Journal of Solids and Structures*, 87: 11–25, 2016.

11 Aleksey D Drozdov. Swelling of pH-responsive cationic gels: constitutive modeling and structure–property relations. *International Journal of Solids and Structures*, 64: 176–190, 2015.

12 Jianyu Li, Zhigang Suo, and Joost J Vlassak. A model of ideal elastomeric gels for polyelectrolyte gels. *Soft Matter*, 10(15): 2582–2590, 2014.

13 Shawn A Chester and Lallit Anand. A thermo-mechanically coupled theory for fluid permeation in elastomeric materials: application to thermally responsive gels. *Journal of the Mechanics and Physics of Solids*, 59(10): 1978–2006, 2011.

14 Aishwarya Pantula, Bibekananda Datta, Yupin Shi, Margaret Wang, Jiayu Liu, Siming Deng, Noah J Cowan, Thao D Nguyen, and David H Gracias. Untethered unidirectionally crawling gels driven by asymmetry in contact forces. *Science Robotics*, 7(73): eadd2903, 2022.

15 Sooraj Narayan and Lallit Anand. A coupled electro-chemo-mechanical theory for polyelectrolyte gels with application to modeling their chemical stimuli-driven swelling response. *Journal of the Mechanics and Physics of Solids*, 159: 104734, 2022.

16 Brandon K Zimmerman, Bibekananda Datta, Ruohong Shi, Rebecca Schulman, and Thao D Nguyen. A reactive electrochemomechanical theory for growth and remodeling of polyelectrolyte hydrogels and application to dynamic polymerization of dna hydrogels. *Journal of the Mechanics and Physics of Solids*, 186: 105568, 2024.

17 A Sydney Gladman, Elisabetta A Matsumoto, Ralph G Nuzzo, Lakshminarayanan Mahadevan, and Jennifer A Lewis. Biomimetic 4D printing. *Nature Materials*, 15(4): 413–418, 2016.

18 T Christian Gasser, Ray W Ogden, and Gerhard A Holzapfel. Hyperelastic modelling of arterial layers with distributed collagen fibre orientations. *Journal of The Royal Society Interface*, 3(6): 15–35, 2006.

19 Suman Jaiswal, Rama Hannineh, Siva Nadimpalli, Samuel Lieber, and Shawn A Chester. Characterization and modeling of the in-plane collagen fiber distribution in the porcine dermis. *Medical Engineering & Physics*, 115: 103973, 2023.

20 Yin Liu, Hongwu Zhang, Jiayong Zhang, and Yonggang Zheng. Constitutive modeling for polymer hydrogels: a new perspective and applications to anisotropic hydrogels in free swelling. *European Journal of Mechanics - A/Solids*, 54: 171–186, 2015b.

21 Nikola Bosnjak, Shuolun Wang, Daehoon Han, Howon Lee, and Shawn A Chester. Modeling of fiber-reinforced polymeric gels. *Mechanics Research Communications*, 96: 7–18, 2019.

22 Zhe Chen, Ruishen Lou, Danming Zhong, Rui Xiao, Shaoxing Qu, and Wei Yang. An anisotropic constitutive model for 3D printed hydrogel-fiber composites. *Journal of the Mechanics and Physics of Solids*, 156: 104611, 2021.

23 G Unnikrishnan and Sabu Thomas. Diffusion and transport of aromatic hydrocarbons through natural rubber. *Polymer*, 35(25): 5504–5510, 1994.

24 D De Kee, Q Liu, and J Hinestroza. Viscoelastic (non-Fickian) diffusion. *The Canadian Journal of Chemical Engineering*, 83(6): 913–929, 2005.

25 Sanjay Govindjee and Juan C Simo. Coupled stress-diffusion: case II. *Journal of the Mechanics and Physics of Solids*, 41(5): 863–887, 1993.

26 Tiezheng Qian and Philip L Taylor. From the Thomas–Windle model to a phenomenological description of case-II diffusion in polymers. *Polymer*, 41(19): 7159–7163, 2000.

27 D Vesely. Diffusion of liquids in polymers. *International Materials Reviews*, 53(5): 299–315, 2008.

28 Jana Wilmers and Swantje Bargmann. A continuum mechanical model for the description of solvent induced swelling in polymeric glasses: thermomechanics coupled with diffusion. *European Journal of Mechanics - A/Solids*, 53: 10–18, 2015.

29 Swantje Bargmann, Andrew T McBride, and Paul Steinmann. Models of solvent penetration in glassy polymers with an emphasis on case II diffusion. A comparative review. *Applied Mechanics Reviews*, 64(1): 010803, 2011.

30 Noreen L Thomas and Alan H Windle. A deformation model for case II diffusion. *Polymer*, 21(6): 613–619, 1980.

31 G Rossi, P A Pincus, and P-G De Gennes. A phenomenological description of case-II diffusion in polymeric materials. *Europhysics Letters (EPL)*, 32(5): 391–396, 1995.

32 Prashanth K Vijalapura and Sanjay Govindjee. Numerical simulation of coupled-stress case II diffusion in one dimension. *Journal of Polymer Science Part B: Polymer Physics*, 41(18): 2091–2108, 2003.

33 Zhaoqiang Song, Xu Li, Kaijin Wu, and Shengqiang Cai. Nonequilibrium thermodynamic modeling of case II diffusion in glassy polymers. *Journal of the Mechanics and Physics of Solids*, 179: 105395, 2023.

34 Morton E Gurtin, Eliot Fried, and Lallit Anand. *The Mechanics and Thermodynamics of Continua*. Cambridge University Press, 2010.

35 Lu Dai and Rui Xiao. A thermodynamic-consistent model for the thermo-chemo-mechanical couplings in amorphous shape-memory polymers. *International Journal of Applied Mechanics*, 13(02): 2150022, 2021.

36 Wei Hong, Zishun Liu, and Zhigang Suo. Inhomogeneous swelling of a gel in equilibrium with a solvent and mechanical load. *International Journal of Solids and Structures*, 46(17): 3282–3289, 2009.

37 Shawn A Chester and Lallit Anand. A coupled theory of fluid permeation and large deformations for elastomeric materials. *Journal of the Mechanics and Physics of Solids*, 58(11): 1879–1906, 2010.

38 Paul J Flory. Thermodynamics of high polymer solutions. *The Journal of Chemical Physics*, 10(1): 51–61, 1942.

39 Maurice L Huggins. Some properties of solutions of long-chain compounds. *The Journal of Physical Chemistry*, 46(1): 151–158, 1942.

40 Gerhard A Holzapfel, Thomas C Gasser, and Ray W Ogden. A new constitutive framework for arterial wall mechanics and a comparative study of material models. *Journal of Elasticity and the Physical Science of Solids*, 61: 1–48, 2000.

41 Theresa K Tonge, Liming M Voo, and Thao D Nguyen. Full-field bulge test for planar anisotropic tissues: part II–a thin shell method for determining material parameters and comparison of two distributed fiber modeling approaches. *Acta Biomaterialia*, 9(4): 5926–5942, 2013.

42 Richard M Christensen. *Theory of Viscoelasticity*. Courier Corporation, 2013.

43 Danming Zhong, Yuhai Xiang, Zhicheng Wang, Zhe Chen, Junjie Liu, Zi Liang Wu, Rui Xiao, Shaoxing Qu, and Wei Yang. A visco-hyperelastic model for hydrogels with tunable water content. *Journal of the Mechanics and Physics of Solids*, 173: 105206, 2023.

44 Rui Xiao and Thao D Nguyen. Modeling the solvent-induced shape-memory behavior of glassy polymers. *Soft Matter*, 9(39): 9455–9464, 2013.

45 Rui Xiao and Hai Li. Glass transition in gels. *Physical Review Materials*, 5(6): 065604, 2021.

46 Lu Dai, Junwei Xu, and Rui Xiao. Modeling the stimulus-responsive behaviors of fiber-reinforced soft materials. *International Journal of Applied Mechanics*, 16(06): 2450041, 2024.

47 Shawn A Chester, Claudio V Di Leo, and Lallit Anand. A finite element implementation of a coupled diffusion-deformation theory for elastomeric gels. *International Journal of Solids and Structures*, 52: 1–18, 2015.

48 Junwei Xu, Zhe Chen, Rui Xiao, and Shaoxing Qu. Modeling of transient diffusion in 3d printed fiber-reinforced hydrogel structures. *International Journal of Solids and Structures*, 311: 113248, 2025.

49 Ronald C Lasky, Edward J Kramer, and Chung Yuen Hui. The initial stages of case II diffusion at low penetrant activities. *Polymer*, 29(4): 673–679, 1988.

Index

Note: Page numbers in *italics* and **bold** refers to figures and tables respectively.

a

Adam–Gibbs model 67–68
angular momentum
  balance of 12–13, 126–127
  law of conservation 13

b

balance of angular momentum 12–13, 126–127
balance of energy 9, 14–15, 127–128
balance of mechanical energy 13–14
balance of momentum 9, 11–12, 82, 146, 152
balance principles 9
  balance of angular momentum 12–13
  balance of energy 14–15
  balance of mechanical energy 13–14
  balance of momentum 11–12
  conservation of mass 11
  entropy inequality 15–17
  material derivative 9
  spatial derivative 9
Biot chain model 28–30, *31*

c

Cauchy deformation tensor 33, 59
Cauchy stress 9, 22, 24–25, 27–28, 34, 43, 65, 147
  nonequilibrium 40, 153
  tensor 13
  viscoelastic model based on 37–38, *38*, *39*
chemomechanical coupled theory 143
classic balance principles 9
Clausius–Planck inequality 16–17
Coleman–Noll argument 130–131
compressible hyperelastic models 23–24
conservation of mass 9, 11
constitutive equations 128–130
  compressible hyperelastic models 23–24
  of hyperelastic materials 21
  incompressible hyperelastic models 22–23
constitutive models 19, 58, **69**, 80, 100, 102, 144
  for dynamic cross-linked networks 20
  hyperelastic 25
  macroscopic 137
  for M-SMPs 106
  physical-based 35
  reduced version 67–68
  visco-hyperelastic-electro 86
constitutive relation 34, 58, 90–91
  diffusion in glassy elastomers 150–152
  fiber-reinforced gels 149–150
  in hydrogels 148
  neutral gels 148–149
  numerical method 152–153
constitutive theory 106–109
continuum damage model 41–42
  for filled rubbers 20

d

damage models 19, 41
  classic 20
  continuum 41–42
  extended network alteration theory 43–45
  extended progressively 45–47
  network alteration theory 42
  progressively 42
  results 47–51, *48–52*, **50–51**
deformation gradient 5–7, *6*, 21, 23, 45, 59, 81, 101, 145, 149
  tensor 6, 9
  time derivative 7

DEs. *see* dielectric elastomers (DEs)
dielectric elastomers (DEs) 79
  constitutive relation 90–91
  electro-hyperelastic model 84–85, *85*
  electro-hyper-viscoelastic model 86
  electromechanical instability of 86–90, *87, 90*
  finite element implementation 90–94
  simulation examples 92–94, *92–94*
  theory 80–90
  thermodynamic framework 81–84
  weak forms 91–94
diffusion
  Fickian 156
  in glassy elastomers 150–152
  in glassy polymers 155–157, *156*
  molecular chain 20
  transient network model 35–36
direct ink writing (DIW) 3D printing 144
divergence 4–5, 7, 127–128
Doolittle equation 151

e

effective temperature theory 57, 67, 71–73, *72, 73*
  performance 71–73
  thermodynamic framework for 59
eight-chain model 19–20, 26–27, 46–47, 64
electric quantities 81
electro-hyperelastic model 84–85, *85*
electro-hyper-viscoelastic model 86
electromechanical instability (EMI) 79–80, 86
  of dielectric elastomers 86–90, *87, 90*
EMI. *see* electromechanical instability (EMI)
energy
  balance of 127–128
  imbalance 82–84
entropy flux 15, 60, 107
entropy inequality 10, 15–17, 34–35, 63, 106
Euler–Almansi strain tensor 8
extended eight-chain model 19, 27–28, *27*, 30, *31*
extended network alteration theory 43–45
extended progressively damage model 45–48

f

F-based model 102–103, *103*
  comparing R-based models and 104–105, *105*
fiber-reinforced gels 149–150, 154–155, *154–155*
finite deformation viscoelastic models 19, 33
  Cauchy stress, viscoelastic model based on 37–38, *38*
  diffusion-based transient network model 35–36
  multiplicative decomposition model 33–34
  second Piola–Kirchhoff Stress, viscoelastic model based on 36–37
finite element
first law of thermodynamics 14, 60, 106–107, 109
free energy density 27–28, 43, 84, 132–134, *134*, 150
  Helmholtz 16–17, 33, 35
  mixing 149
  volumetric 24

g

Gauss's Law 82–83, 91
gels 19–21, 41, 144
glassy elastomers, diffusion in 150–152
glassy polymers 144
  diffusion in 155–157, *156*
gradient 4–5, 130
Green–Lagrange strain tensor 8, 81

h

hard magnetic iron-particle-filled magnetorheological elastomers (h-MREs) 100
  magnetic-mechanical coupled model for 101–105
hard magnetic particles 99–100, *100*
Helmholtz energy density 107
Helmholtz free energy density 16–17, 33, 60–61, 64, 67
  nominal 102
  rate of 33, 35, 107
Helmholtz free energy function 21
h-MREs. *see* hard magnetic iron-particle-filled magnetorheological elastomers (h-MREs)
hydrogels 143
  constitutive relations 148–153
  diffusion in glassy polymers 155–157, *156*
  fiber-reinforced gels 154–155, *154–155*
  neutral gels 154–155, *154–155*
  results 154
  thermodynamics 145–148
hyperelastic models 19–21, 24
  Biot chain model 27–29
  constitutive equations of hyperelastic materials 21–24
  eight-chain model 26–27
  extended eight-chain model 27–28, *27*
  Gent model 26
  Mooney–Rivlin model 25–26
  Neo-Hookean model 25
  Ogden model 26
  results 29–31, *30, 31*

i

index notation 2–3
infinitesimal volume element 7
  deformation state of 6

integral theorem 15, 145
isochoric Cauchy–Green deformation tensor 23
isochoric second Piola–Kirchhoff stress 24, 36

k

kinematics 5
  deformation gradient 5–7, *6*
  of dielectric elastomers 81
  of shape-memory polymers 59
  strain tensor 7–8
Kronecker delta 3

l

Laplacian operator 5
law of conservation 13
  of angular momentum 13
  of momentum 12
LCEs. *see* liquid crystal elastomers (LCEs)
left Cauthy–Green tensor 8
linear momentum, balance of 127–128
liquid crystal 119–121, *119*
liquid crystal elastomers (LCEs) 117
  general theory for 119
  liquid crystal 119–121, *119*
  nematic polymers 121–123
  theory for polydomain LCEs 134–139, *135, 137–139*
  thermomechanical coupled model for monodomain LCEs 123–125
  viscoelastic micropolar theory for monodomain LCEs 125–134, *126*

m

magnetic-activated shape-memory polymers 105, *106*
  constitutive theory 106–109
  parameter determination 109–111, *111*, **112**
  simulation results 111–112, *112*
magnetic-mechanical coupled model
  comparing R-based and F-based models 104–105, *105*
  F-based model 102–103, *103*
  for h-MREs 101
  kinematics 101
  R-based model 103–104
magnetoactive soft materials 99, *100*
  magnetic-activated shape-memory polymers 105–112, *106*
  magnetic-mechanical coupled model for h-MREs 101–105
magnetorheological elastomers (MREs) 99
Mandel stress 34, 63, 65
material derivative 10, 127
material isotropy 22
micromechancial models 24
micropolar continua theory 126
monodomain LCEs 117
  thermomechanical coupled model for 123–125, *124–125*
  viscoelastic micropolar theory for 125–134, *126*
Mooney–Rivlin model 25–26, 123
MREs. *see* magnetorheological elastomers (MREs)
Mullins effect 20, 42, 49
multiplicative decomposition model 33–34, 40

n

Nanson's equation 7, 9
nematic polymers 121
  neoclassical model 122–123, *123*
  step length tensor 121–122
nematic-isotropic (N-I) phase transition 118
neoclassical model 118, 122–123, *123*
Neo-Hookean Model 25, 29, 34, 37, 41, 80, 88, 90
  for mechanical free energy 84
  quasi-incompressible 103
network alteration theory 42
network deformation gradient 132
neutral gels 144, 148–149, 154–155, *154–155*
nominal solvent number 147
nonequilibrium moduli 110
non-gradient case, simplification for 130–132
numerical method 152–153

o

Ogden model 20, 26
one-dimensional small strain viscoelastic models 31–32, *32*

p

permutation symbol 3
phenomenological models 19–20, 24
Piola–Kirchhoff stress tensor 9, 21, 60, 65, 82, 106
polar decomposition theorem 21
polydomain liquid crystal elastomers, theory for 134–139, *135, 137–139*
polydomain-to-monodomain (P-M) transition 118
polyelectrolyte gels 144
poly(ethylene terephthalate)-glycol (PETG) 68
polymer ratio 145
progressively damage model 42

r

R-based model 103–104
  comparing F-based models and 104–105, *105*
reference electric field 81
relaxation time 32, 34, 36–38, 67–68, 86, 92, 110, 151
Reynolds Transport Theorem 10–13

right Cauchy–Green tensor 8
  deviatoric 90
  function of 83

s

second law of thermodynamics 9, 15, 60–61
  balance of 127–128
second-order tensor 3–4
second Piola–Kirchhoff stress tensor 65
  viscoelastic model based on 36–37
shape-memory polymers (SMPs) 57
  constitutive model, reduced version of 67–68
  constitutive relationships 64–67
  effective temperature theory, performance of 71–73, *72–73*
  kinematics 59
  parameter determination 68
  results 68
  thermodynamic framework 60–64
  thermodynamics 59–68
  viscoelastic models, performance of 68–71, **69**, **70**, *70–71*
shear loss modulus 69
simulation examples 92–94, *92–94*
SMPs. *see* shape-memory polymers (SMPs)
s-MREs. *see* soft-magnetic-particle-filled magnetorheological elastomers (s-MREs)
soft-magnetic-particle-filled magnetorheological elastomers (s-MREs) 99
soft magnetic particles 99, 105, 113
soft materials 19, 102
soft polymeric materials 21
solvent-induced shape-memory effects 58
spatial derivative 10
step length tensor 121–122
strain tensor 7–8
stress 8–9

t

temperature memory effect 58
tensor 36
thermodynamics
  of hydrogels 145–148
  of shape-memory polymers 59–68
thermomechanical coupled model
  for monodomain liquid crystal elastomers 123–125
thermo-responsive gels 144

u

unit vector tensor 3

v

vector 1–2
vector operator 4
velocity gradients 59
VHB elastomers 80, 86
viscoelastic micropolar theory
  balance of energy and second law of thermodynamics 127–128
  balance of linear and angular momentum 126–127
viscoelastic micropolar theory
  constitutive equations 128–130
  free energy density 132–134, *134*
  kinematics 126
  for monodomain liquid crystal elastomers 125–134, *126*
  simplification for non-gradient case 130–132
viscoelastic model 20, 31, 69, 71
  based on Cauchy stress 37–38, *38, 39*
  based on second Piola–Kirchhoff stress 36–37
  finite deformation 33–38
  finite element simulation 39–41, *41*
  one-dimensional small strain 31–32, *32*
  performance of 68–71, **69**, **70**, *70–71*
  results 38–39, *39, 40*
viscoelastic relaxation spectra 110
viscoelastic stress 37
viscosity-type parameter 66
viscous deformation process 150

w

weak forms 91–94
Williams–Landel–Ferry (WLF) equation 66, 110
WLF. *see* Williams–Landel–Ferry (WLF) equation